Thomas Henry Huxley

Social diseases and worse remedies

Thomas Henry Huxley

Social diseases and worse remedies

ISBN/EAN: 9783337109349

Printed in Europe, USA, Canada, Australia, Japan

Cover: Foto ©berggeist007 / pixelio.de

More available books at **www.hansebooks.com**

SOCIAL DISEASES
AND WORSE REMEDIES:

LETTERS TO THE "TIMES" ON MR. BOOTH'S SCHEME

WITH A

PREFACE AND (reprinted) INTRODUCTORY ESSAY

BY

T. H. HUXLEY, F.R.S.

"*Sixpennyworth of good and a shilling's worth of harm*"

SECOND EDITION—FOURTH THOUSAND

London
MACMILLAN AND CO.
AND NEW YORK
1891

The Right of Translation and Reproduction is Reserved

PREFACE

THE letters which are here collected together were published in the *Times* in the course of the months of December, 1890, and January, 1891.

The circumstances which led me to write the first letter are sufficiently set forth in its opening sentences; and the materials on which I based my criticisms of Mr. Booth's scheme, in this and in the second letter, were wholly derived from Mr. Booth's book. I had some reason to know, however, that when anybody allows his sense of duty so far to prevail over his sense of the blessedness of peace as to write a letter to the *Times*, on any subject of public interest, his reflections, before he has done with the business, will be very like those of Johnny Gilpin, "who little thought, when he set out, of running such a rig." Such undoubtedly are mine when I contemplate these ten documents, and call to mind the distinct addition to the revenue of the Post Office which must have accrued from the mass of letters and pamphlets which have been delivered at my door; to say nothing of the unexpected light upon my character, motives, and doctrines, which has been thrown by some of the *Times*' correspondents, and by no end of comments elsewhere.

If self-knowledge is the highest aim of man, I ought by this time to have little to learn. And yet, if I am awake, some of my teachers—unable, perhaps, to control the divine fire of the poetic imagination which is so closely akin to, if not a part of, the mythopœic faculty—have surely dreamed dreams. So far as my humbler and essentially prosaic faculties of observation and comparison go, plain facts are against them. But, as I may be mistaken, I have thought it well to prefix to the letters (by way of "Prolegomena") an essay which appeared in the *Nineteenth Century* for January, 1888, in which the principles that, to my mind, lie at the bottom of the "social question" are stated. So far as Individualism and Regimental Socialism are concerned, this paper simply emphazises and expands the opinions expressed in an address to the members of the Midland Institute, delivered seventeen years earlier, and still more fully developed in several essays published in the *Nineteenth Century* in 1889, which I hope, before long, to republish.

The fundamental proposition which runs through the writings, which thus extend over a period of twenty years, is, that the common *a priori* doctrines and methods of reasoning about political and social questions are essentially vicious; and that argumentation on this basis leads, with equal logical force, to two contradictory and extremely mischievous systems, the one that of Anarchic Individualism, the other that of despotic or Regimental Socialism. Whether I am right or wrong, I am at least consistent in opposing both to the best of my ability. Mr. Booth's system appears to me, and, as I have shown, is regarded by Socialists themselves, to be mere autocratic Socialism, masked by its theological exterior. That the "fantastic" religious skin will wear away, and the Socialistic reality it covers will show its real nature, is the expressed hope of one candid Socialist, and may be fairly conceived to be the unexpressed belief of the despotic leader of the new Trades Union, who has shown his zeal, if not his discretion, in championing Mr. Booth's projects.

Yet another word to commentators upon my letters. There are some who rather chuckle, and some who sneer, at what they seem to consider the dexterity of an "old controversial hand," exhibited by the contrast which I have drawn between the methods of conversion depicted in the New Testament and those pursued by fanatics of the Salvationist type, whether they be such as are now exploited by Mr. Booth, or such as those who, from the time of the Anabaptists, to go no further back, have worked upon similar lines.

Whether such observations were intended to be flattering or sarcastic, I must respectfully decline to accept the compliment or to apply the sarcasm to myself. I object to obliquity of procedure and ambiguity of speech in all shapes. And I confess that I find it difficult to understand the state of mind which leads any one to suppose, that deep respect for single-minded devotion to high aims is incompatible with the unhesitating conviction that those aims include the propagation of doctrines which are devoid of foundation—perhaps even mischievous.

The most degrading feature of the narrower forms of Christianity (of which that professed by Mr. Booth is a notable example) is their insistence that the noblest virtues, if displayed by those who reject their pitiable formulæ, are, as their pet phrase goes, "splendid sins." But there is, perhaps, one step lower; and that is that men, who profess freedom of thought, should fail to see and appreciate that large soul of goodness which often animates even the fanatical adherents of such tenets. I am sorry for any man who can read the epistles to the Galatians and the Corinthians without yielding a large meed of admiration to the fervent humanity of Paul of Tarsus; who can study the lives of Francis of Assisi, or of Catherine of Siena, without wishing that, for the furtherance of his own ideals, he might be even as they; or who can contemplate unmoved the steadfast veracity and true heroism which loom through the fogs of mystical utterance in George Fox. In all these great men and women there lay the root of the matter—a burning desire to amend the condition of their fellow-men, and to put aside all other things for that end. If, in spite of all the dogmatic helps or hindrances in which they were entangled, these people are not to be held in high honour, who are?

I have never expressed a doubt—for I have none—that, when Mr. Booth left the Methodist connection, and started that organization of the Salvation Army upon which, comparatively recently, such ambitious schemes of social reform have been grafted, he may have deserved some share of such honour. I do not say that, so far as his personal desires and intentions go, he may not still deserve it.

But the correlate of despotic authority is unlimited responsibility. If Mr. Booth is to take credit for any good that the Army system has effected, he must be prepared to bear blame for its inherent evils. As it seems to me, that has happened to him which sooner or later happens to all despots : he has become the slave of his own creation—the prosperity and glory of the soul-saving machine have become the end, instead of a means, of soul-saving; and to maintain these at the proper pitch, the "General" is led to do things which the Mr. Booth of twenty years ago would probably have scorned.

And those who desire, as I most emphatically desire, to be just to Mr. Booth, however badly they may think of the working of the organization he has founded, will bear in mind that some astute backers of his probably care little enough for Salvationist religion ; and are perhaps not very keen about many of Mr. Booth's projects. I have referred to the rubbing of the hands of the Socialists over Mr. Booth's success ;[1] but, unless I err greatly, there are politicians of a certain school to whom it affords still greater satisfaction. Consider what electioneering agents the

[1] See p. 103.

captains of the Salvation Army, scattered through all our towns, and directed from a political "bureau" in London, would make! Think how political adversaries could be harassed by our local attorney—"tribune of the people" I mean; and how a troublesome man, on the other side, could be "hunted down" upon any convenient charge, whether true or false, brought by our Vigilance-familiar![1]

I entirely acquit Mr. Booth of any complicity in far-reaching schemes of this kind; but I did not write idly when, in my first letter, I gave no vague warning of what might grow out of the organised force, drilled in the habit of unhesitating obedience, which he has created.

T. H. H.

[1] See pp. 66, 67.

SOCIAL DISEASES AND WORSE REMEDIES

THE STRUGGLE FOR EXISTENCE IN HUMAN SOCIETY.

THE vast and varied procession of events, which we call Nature, affords a sublime spectacle and an inexhaustible wealth of attractive problems to the speculative observer. If we confine our attention to that aspect which engages the attention of the intellect, nature appears a beautiful and harmonious whole, the incarnation of a faultless logical process, from certain premises in the past to an inevitable conclusion in the future. But if it be regarded from a less elevated, though more human, point of view; if our moral sympathies are allowed to influence our judgment, and we permit ourselves to criticize our great mother as we criticize one another;—then our verdict, at least so far as sentient nature is concerned, can hardly be so favourable.

In sober truth, to those who have made a study of the phenomena of life as they are exhibited by the higher forms of the animal world, the optimistic dogma, that this is the best of all possible worlds, will seem little better than a libel upon possibility. It is really only another instance to be added to the many extant, of the audacity of *a priori* speculators who, having created God in their own image, find no difficulty in assuming that the Almighty must have been actuated by the same motives as themselves. They are quite sure that, had any other course been practicable, He would no more have made infinite suffering a necessary ingredient of His handiwork than a respectable philosopher would have done the like.

But even the modified optimism of the time-honoured thesis of physico-theology, that the sentient world is, on the whole, regulated by principles of benevolence, does but ill stand the test of impartial confrontation with the facts of the case. No doubt it is quite true that sentient nature affords hosts of examples of subtle contrivances directed towards the production of pleasure or the avoidance of pain ; and it may be proper to say that these are evidences of benevolence. But if so, why is it not equally proper to say of the equally numerous arrangements, the no less necessary result of which is the production of pain, that they are evidences of malevolence?

If a vast amount of that which, in a piece of

human workmanship, we should call skill, is visible in those parts of the organization of a deer to which it owes its ability to escape from beasts of prey, there is at least equal skill displayed in that bodily mechanism of the wolf which enables him to track, and sooner or later to bring down, the deer. Viewed under the dry light of science, deer and wolf are alike admirable; and if both were non-sentient automata, there would be nothing to qualify our admiration of the action of the one on the other. But the fact that the deer suffers, while the wolf inflicts suffering, engages our moral sympathies. We should call men like the deer innocent and good, men such as the wolf malignant and bad; we should call those who defended the deer and aided him to escape brave and compassionate, and those who helped the wolf in his bloody work base and cruel. Surely, if we transfer these judgments to nature outside the world of man at all, we must do so impartially. In that case, the goodness of the right hand which helps the deer, and the wickedness of the left hand which eggs on the wolf, will neutralize one another: and the course of nature will appear to be neither moral nor immoral, but non-moral.

This conclusion is thrust upon us by analogous facts in every part of the sentient world; yet, inasmuch as it not only jars upon prevalent prejudices, but arouses the natural dislike to that which is painful, much ingenuity has been exercised in devising an escape from it.

From the theological side, we are told that this is a state of probation, and that the seeming injustices and immoralities of nature will be compensated by and by. But how this compensation is to be effected, in the case of the great majority of sentient things, is not clear. I apprehend that no one is seriously prepared to maintain that the ghosts of all the myriads of generations of herbivorous animals which lived during the millions of years of the earth's duration, before the appearance of man, and which have all that time been tormented and devoured by carnivores, are to be compensated by a perennial existence in clover; while the ghosts of carnivores are to go to some kennel where there is neither a pan of water nor a bone with any meat on it. Besides, from the point of view of morality, the last state of things would be worse than the first. For the carnivores, however brutal and sanguinary, have only done that which, if there is any evidence of contrivance in the world, they were expressly constructed to do. Moreover carnivores and herbivores alike have been subject to all the miseries incidental to old age, disease, and over-multiplication, and both might well put in a claim for "compensation" on this score.

On the evolutionist side, on the other hand, we are told to take comfort from the reflection that the terrible struggle for existence tends to final good, and that the suffering of the ancestor is paid for by the increased perfection of the progeny. There would be something in this argument if, in Chinese fashion, the present generation could pay its debts to its ancestors; otherwise it is not clear what compensation the *Eohippus* gets for his sorrows in the fact that, some millions of years afterwards, one of his descendants wins the Derby. And, again, it is an error to imagine that evolution signifies a constant tendency to increased perfection. That process undoubtedly involves a constant re-modelling of the organism in adaptation to new conditions; but it depends on the nature of those conditions whether the direction of the modifications effected shall be upward or downward. Retrogressive is as practicable as progressive metamorphosis. If what the physical philosophers tell us, that our globe has been in a state of fusion, and, like the sun, is gradually cooling down, is true; then the time must come when evolution will mean adaptation to an universal winter, and all forms of life will die out, except such low and simple organisms as the Diatom of the arctic and antarctic ice and the Protococcus of the red snow. If our globe is proceeding from a condition in which it was too hot to support any but the lowest living thing to a condition in which it will be too cold to permit of the existence of any others, the course of life upon its surface must describe a trajectory like that of a ball fired from a mortar; and the sinking half of that course is as much a part of the general process of evolution as the rising.

From the point of view of the moralist the animal world is on about the same level as a gladiator's show.

The creatures are fairly well treated, and set to fight —whereby the strongest, the swiftest and the cunningest live to fight another day. The spectator has no need to turn his thumbs down, as no quarter is given. He must admit that the skill and training displayed are wonderful. But he must shut his eyes if he would not see that more or less enduring suffering is the meed of both vanquished and victor. And since the great game is going on in every corner of the world, thousands of times a minute; since, were our ears sharp enough, we need not descend to the gates of hell to hear—

> sospiri, pianti, ed alti guai.
> . . . .
> Voci alte e fioche, e suon di man con elle.

It seems to follow that, if this world is governed by benevolence, it must be a different sort of benevolence from that of John Howard.

But the old Babylonians wisely symbolized Nature by their great goddess Istar, who combined the attributes of Aphrodite with those of Ares. Her terrible aspect is not to be ignored or covered up with shams; but it is not the only one. If the optimism of Leibnitz is a foolish though pleasant dream, the pessimism of Schopenhauer is a nightmare, the more foolish because of its hideousness. Error which is not pleasant is surely the worst form of wrong.

This may not be the best of all possible worlds, but to say that it is the worst is mere petulant nonsense. A worn-out voluptuary may find nothing good under the sun, or a vain and inexperienced youth, who cannot get the moon he cries for, may vent his irritation in pessimistic moanings; but there can be no doubt in the mind of any reasonable person that mankind could, would, and in fact do, get on fairly well with vastly less happiness and far more misery than find their way into the lives of nine people out of ten. If each and all of us had been visited by an attack of neuralgia, or of extreme mental depression, for one hour in every twenty-four—a supposition which many tolerably vigorous people know, to their cost, is not extravagant—the burden of life would have been immensely increased without much practical hindrance to its general course. Men with any manhood in them find life quite worth living under worse conditions than these.

There is another sufficiently obvious fact, which renders the hypothesis that the course of sentient nature is dictated by malevolence quite untenable. A vast multitude of pleasures, and these among the purest and the best, are superfluities, bits of good which are to all appearance unnecessary as inducements to live, and are, so to speak, thrown into the bargain of life. To those who experience them, few delights can be more entrancing than such as are afforded by natural beauty or by the arts and especially by music; but they are products of, rather than factors in, evolution, and it is probable that they are known, in any considerable degree, to but a very small proportion of mankind.

The conclusion of the whole matter seems to be that, if Ormuzd has not had his way in this world, neither has Ahriman. Pessimism is as little consonant with the facts of sentient existence as optimism. If we desire to represent the course of nature in terms of human thought, and assume that it was intended to be that which it is, we must say that its governing principle is intellectual and not moral; that it is a materialized logical process accompanied by pleasures and pains, the incidence of which, in the majority of cases, has not the slightest reference to moral desert. That the rain falls alike upon the just and the unjust, and that those upon whom the Tower of Siloam fell were no worse than their neighbours, seem to be Oriental modes of expressing the same conclusion.

In the strict sense of the word "nature," it denotes the sum of the phenomenal world, of that which has been, and is, and will be; and society, like art, is therefore a part of nature. But it is convenient to distinguish those parts of nature in which man plays the part of immediate cause, as something apart; and, therefore, society, like art, is usefully to be considered as distinct from nature. It is the more desirable, and even necessary, to make this distinction, since society differs from nature in having a definite moral object; whence it comes about that the course shaped by the ethical man—the member of society or citizen—necessarily runs counter to that which the non-ethical man—the primitive savage, or man as a mere member of the animal kingdom—tends to adopt. The latter fights out the struggle for existence to the bitter end, like any other animal; the former devotes his best energies to the object of setting limits to the struggle.

In the cycle of phenomena presented by the life of man, the animal, no more moral end is discernible than in that presented by the lives of the wolf and of the deer. However imperfect the relics of prehistoric men may be, the evidence which they afford clearly tends to the conclusion that, for thousands and thousands of years, before the origin of the oldest known civilizations, men were savages of a very low type. They strove with their enemies and their competitors; they preyed upon things weaker or less cunning than themselves; they were born, multiplied without stint, and died, for thousands of generations alongside the mammoth, the urus, the lion, and the hyæna, whose lives were spent in the same way; and they were no more to be praised or blamed, on moral grounds, than their less erect and more hairy compatriots.

As among these, so among primitive men, the weakest and stupidest went to the wall, while the toughest and shrewdest, those who were best fitted to cope with their circumstances, but not the best in any other sense, survived. Life was a continual free fight, and beyond the limited and temporary relations of the family, the Hobbesian war of each against all was the normal state of existence. The human species, like others, plashed and floundered amid the general stream of evolution, keeping its head above water as it best might, and thinking neither of whence nor whither.

The history of civilization—that is of society—on the other hand, is the record of the attempts which the human race has made to escape from this position. The first men who substituted the state of mutual peace for that of mutual war, whatever the motive which impelled them to take that step, created society. But, in establishing peace, they obviously put a limit upon the struggle for existence. Between the members of that society, at any rate, it was not to be pursued *à outrance.* And of all the successive shapes which society has taken, that most nearly approaches perfection in which the war of individual against individual is most strictly limited. The primitive savage, tutored by Istar, appropriated whatever took his fancy, and killed whomsoever opposed him, if he could. On the contrary, the ideal of the ethical man is to limit his freedom of action to a sphere in which he does not interfere with the freedom of others; he seeks the common weal as much as his own; and, indeed, as an essential part of his own welfare. Peace is both end and means with him; and he founds his life on a more or less complete self-restraint, which is the negation of the unlimited struggle for existence. He tries to escape from his place in the animal kingdom, founded on the free development of the principle of non-moral evolution, and to establish a kingdom of Man, governed upon the principle of moral evolution. For society not only has a moral end, but in its perfection, social life, is embodied morality.

But the effort of ethical man to work towards a moral end by no means abolished, perhaps has hardly modified, the deep-seated organic impulses which impel the natural man to follow his non-moral course. One of the most essential conditions, if not the chief cause, of the struggle for existence, is the tendency to multiply without limit, which man shares with all living things. It is notable that "increase and multiply" is a commandment traditionally much older than the ten ; and that it is, perhaps, the only one which has been spontaneously and *ex animo* obeyed by the great majority of the human race. But, in civilized society, the inevitable result of such obedience is the re-establishment, in all its intensity, of that struggle for existence—the war of each against all—the mitigation or abolition of which was the chief end of social organization.

It is conceivable that, at some period in the history of the fabled Atlantis, the production of food should have been exactly sufficient to meet the wants of the population, that the makers of the commodities of the artificer should have amounted to just the number supportable by the surplus food of the agriculturists. And, as there is no harm in adding another monstrous supposition to the foregoing, let it be imagined that every man, woman, and child was perfectly virtuous, and aimed at the good of all as the highest personal good. In that happy land, the natural man would have been finally put down by the ethical man. There would have been no competition, but the industry of each would have been serviceable to all; nobody being vain and nobody avaricious, there would have been no rivalries; the struggle for existence would have been abolished, and the millennium would have finally set in. But it is obvious that this state of things could have been permanent only with a stationary population. Add ten fresh mouths; and as, by the supposition, there was only exactly enough before, somebody must go on short rations. The Atlantis society might have been a heaven upon earth, the whole nation might have consisted of just men, needing no repentance, and yet somebody must starve. Reckless Istar, non-moral Nature, would have riven the ethical fabric. I was once talking with a very eminent physician[1] about the *vis medicatrix naturæ*. "Stuff!" said he; "nine times out of ten nature does not want to cure the man; she wants to put him in his coffin." And Istar-Nature appears to have equally little sympathy with the ends of society. "Stuff! she wants nothing but a fair field and free play for her darling the strongest."

Our Atlantis may be an impossible figment, but the antagonistic tendencies which the fable adumbrates have existed in every society which was ever

[1] The late Sir W. Gull.

established, and, to all appearance, must strive for the victory in all that will be. Historians point to the greed and ambition of rulers, to the reckless turbulence of the ruled, to the debasing effects of wealth and luxury, and to the devastating wars which have formed a great part of the occupation of mankind, as the causes of the decay of states and the foundering of old civilizations, and thereby point their story with a moral. No doubt immoral motives of all sorts have figured largely among the minor causes of these events. But, beneath all this superficial turmoil, lay the deep-seated impulse given by unlimited multiplication. In the swarms of colonies thrown out by Phœnicia and by old Greece; in the *ver sacrum* of the Latin races; in the floods of Gauls and of Teutons which burst over the frontiers of the old civilization of Europe; in the swaying to and fro of the vast Mongolian hordes in late times, the population problem comes to the front in a very visible shape. Nor is it less plainly manifest in the everlasting agrarian questions of ancient Rome than in the Arreoi societies of the Polynesian Islands.

In the ancient world and in a large part of that in which we now live, the practice of infanticide was or is a regular and legal custom; the steady recurrence of famine, pestilence, and war, were and are normal factors in the struggle for existence, and have served, in a gross and brutal fashion, to mitigate the intensity of the effects of its chief cause.

But, in the more advanced civilizations, the progress

of private and public morality has steadily tended to remove all these checks. We declare infanticide murder, and punish it as such; we decree, not quite successfully, that no one shall die of hunger; we regard death from preventable causes of other kinds as a sort of constructive murder, and eliminate pestilence to the best of our ability; we declaim against the curse of war, and the wickedness of the military spirit, and we are never weary of dilating on the blessedness of peace and the innocent beneficence of Industry. In their moments of expansion, even statesmen and men of business go thus far. The finer spirits look to an ideal *civitas Dei*: a state when every man, having reached the point of absolute self-negation, and having nothing but moral perfection to strive after, peace will truly reign, not merely among nations, but among men, and the struggle for existence will be at an end.

Whether human nature is competent, under any circumstances, to reach, or even seriously advance towards, this ideal condition, is a question which need not be discussed. It will be admitted that mankind has not yet reached this stage by a very long way, and my business is with the present. And that which I wish to point out is that, so long as the natural man increases and multiplies without restraint, so long will peace and industry not only permit, but they will necessitate, a struggle for existence as sharp as any that ever went on under the *régime* of war. If Istar is to reign on the one hand, she will demand her human sacrifices on the other.

Let us look at home. For seventy years, peace and industry have had their way among us with less interruption and under more favourable conditions than in any other country on the face of the earth. The wealth of Crœsus was nothing to that which we have accumulated, and our prosperity has filled the world with envy. But Nemesis did not forget Crœsus; has she forgotten us?

I think not. There are now 36,000,000 of people in our island, and every year considerably more than 300,000 are added to our numbers.[1] That is to say, about every hundred seconds, or so, a new claimant to a share in the common stock of maintenance presents him or herself among us. At the present time, the produce of the soil does not suffice to feed half its population. The other moiety has to be supplied with food which must be bought from the people of food-producing countries. That is to say, we have to offer them the things which they want in exchange for the things we want. And the things they want and which we can produce better than they can are mainly manufactures—industrial products.

The insolent reproach of the first Napoleon had a very solid foundation. We not only are, but, under

[1] These numbers are only approximately accurate. In 1881, our population amounted to 35,241,482, exceeding the number in 1871 by 3,396,103. The average annual increase in the decennial period 1871—1881 is therefore 339,610. The number of minutes in a calendar year is 525,600.

penalty of starvation, we are bound to be, a nation of shopkeepers. But other nations also lie under the same necessity of keeping shop, and some of them deal in the same goods as ourselves. Our customers naturally seek to get the most and the best in exchange for their produce. If our goods are inferior to those of our competitors, there is no ground, compatible with the sanity of the buyers, which can be alleged, why they should not prefer the latter. And, if that result should ever take place on a large and general scale, five or six millions of us would soon have nothing to eat. We know what the cotton famine was; and we can therefore form some notion of what a dearth of customers would be.

Judged by an ethical standard, nothing can be less satisfactory than the position in which we find ourselves. In a real, though incomplete, degree we have attained the condition of peace which is the main object of social organization ; and it may, for argument's sake, be assumed that we desire nothing but that which is in itself innocent and praiseworthy— namely, the enjoyment of the fruits of honest industry. And lo! in spite of ourselves, we are in reality engaged in an internecine struggle for existence with our presumably no less peaceful and well-meaning neighbours. We seek peace and we do not ensue it. The moral nature in us asks for no more than is compatible with the general good ; the non-moral nature proclaims and acts upon that fine old Scottish family motto " Thou shalt starve ere I want." Let us be under no illusions then. So long as unlimited multiplication goes on, no social organization which has ever been devised, or is likely to be devised; no fiddle-faddling with the distribution of wealth, will deliver society from the tendency to be destroyed by the reproduction within itself, in its intensest form, of that struggle for existence, the limitation of which is the object of society. And however shocking to the moral sense this eternal competition of man against man and of nation against nation may be; however revolting may be the accumulation of misery at the negative pole of society, in contrast with that of monstrous wealth at the positive pole; this state of things must abide, and grow continually worse, so long as Istar holds her way unchecked. It is the true riddle of the Sphinx; and every nation which does not solve it will sooner or later be devoured by the monster itself has generated.

The practical and pressing question for us, just now, seems to me to be how to gain time. "Time brings counsel," as the Teutonic proverb has it; and wiser folk among our posterity may see their way out of that which at present looks like an *impasse.*

It would be folly to entertain any ill-feeling towards those neighbours and rivals who, like ourselves, are slaves of Istar; but, if somebody is to be starved, the modern world has no Oracle of Delphi to which the nations can appeal for an indication of the victim It is open to us to try our fortune; and if we avoid impending fate, there will be a certain ground for believing that we are the right people to escape. *Securus judicat orbis.*

To this end, it is well to look into the necessary conditions of our salvation by works. They are two, one plain to all the world and hardly needing insistence; the other seemingly not so plain, since too often it has been theoretically and practically left out of sight. The obvious condition is that our produce shall be better than that of others. There is only one reason why our goods should be preferred to those of our rivals—our customers must find them better at the price. That means that we must use more knowledge, skill, and industry in producing them, without a proportionate increase in the cost of production; and, as the price of labour constitutes a large element in that cost, the rate of wages must be restricted within certain limits. It is perfectly true that cheap production and cheap labour are by no means synonymous; but it is also true that wages cannot increase beyond a certain proportion without destroying cheapness. Cheapness, then, with, as part and parcel of cheapness, a moderate price of labour, is essential to our success as competitors in the markets of the world.

The second condition is really quite as plainly indispensable as the first, if one thinks seriously about the matter. It is social stability. Society is stable when the wants of its members obtain as much satisfaction as, life being what it is, common sense and experience show may be reasonably expected. Mankind, in general, care very little for forms of government or ideal considerations of any sort; and nothing really stirs the great multitude to break with custom and incur the manifest perils of revolt except the belief that misery in this world or damnation in the next, or both, are threatened by the continuance of the state of things in which they have been brought up. But when they do attain that conviction, society becomes as unstable as a package of dynamite, and a very small matter will produce the explosion which sends it back to the chaos of savagery.

It needs no argument to prove that when the price of labour sinks below a certain point, the worker infallibly falls into that condition which the French emphatically call *la misère*—a word for which I do not think there is any exact English equivalent. It is a condition in which the food, warmth and clothing which are necessary for the mere maintenance of the functions of the body in their normal state cannot be obtained; in which men, women and children are forced to crowd into dens wherein decency is abolished and the most ordinary conditions of healthful existence are impossible of attainment; in which the pleasures within reach are reduced to bestiality and drunkenness; in which the pains accumulate at compound interest, in the shape of starvation, disease, stunted development, and moral degradation; in which the prospect of even steady and honest industry is a life of unsuccessful battling with hunger, rounded by a pauper's grave.

That a certain proportion of the members of every great aggregation of mankind should constantly tend to establish and populate such a Slough of Despond as this is inevitable, so long as some people are by nature idle and vicious, while others are disabled by sickness or accident, or thrown upon the world by the death of their bread-winners. So long as that proportion is restricted within tolerable limits, it can be dealt with; and, so far as it arises only from such causes, its existence may and must be patiently borne. But, when the organization of society, instead of mitigating this tendency, tends to continue and intensify it; when a given social order plainly makes for evil and not for good, men naturally enough begin to think it high time to try a fresh experiment. The animal man, finding that the ethical man has landed him in such a slough, resumes his ancient sovereignty and preaches anarchy; which is, substantially, a proposal to reduce the social cosmos to chaos and begin the brute struggle for existence once again.

Any one who is acquainted with the state of the population of all great industrial centres, whether in this or other countries, is aware that, amidst a large and increasing body of that population, *la misère* reigns supreme. I have no pretensions to the character of a philanthropist and I have a special horror of all sorts of sentimental rhetoric; I am merely trying to deal with facts, to some extent within my own knowledge, and further evidenced by abundant testimony, as a naturalist; and I take it to be a mere plain truth that, throughout industrial Europe, there is not a single large manufacturing city which is free from a vast mass of people whose condition is exactly that described, and from a still greater mass who, living just on the edge of the social swamp, are liable to be precipitated into it by any lack of demand for their produce. And, with every addition to the population, the multitude already sunk in the pit and the number of the host sliding towards it continually increase.

Argumentation can hardly be needful to make it clear that no society, in which the elements of decomposition are thus swiftly and surely accumulating, can hope to win in the race of industries.

Intelligence, knowledge, and skill are undoubtedly conditions of success; but of what avail are they likely to be unless they are backed up by honesty, energy, good-will, and all the physical and moral faculties that go to the making of manhood, and unless they are stimulated by hope of such reward as men may fairly look to? And what dweller in the slough of want, dwarfed in body and soul, demoralized, hopeless, can reasonably be expected to possess these qualities?

Any full and permanent development of the productive powers of an industrial population, then, must be compatible with and, indeed, based upon a social organization which will secure a fair amount of physical and moral welfare to that population ; which will make for good and not for evil. Natural science and religious enthusiasm rarely go hand in hand, but on this matter their concord is complete ; and the least sympathetic of naturalists can but admire the insight and the devotion of such social reformers as the late Lord Shaftesbury, whose recently published *Life and Letters* gives a vivid picture of the condition of the working classes fifty years ago, and of the pit which our industry, ignoring these plain truths, was then digging under its own feet.

There is, perhaps, no more hopeful sign of progress among us in the last half-century than the steadily-increasing devotion which has been and is directed to measures for promoting physical and moral welfare among the poorer classes. Sanitary reformers, like most other reformers whom I have had the advantage of knowing, seem to need a good dose of fanaticism, as a sort of moral coca, to keep them up to the mark, and, doubtless, they have made many mistakes ; but that the endeavour to improve the condition under which our industrial population live, to amend the drainage of densely-peopled streets, to provide baths, washhouses, and gymnasia, to facilitate habits of thrift, to furnish some provision for instruction and amusement in public libraries and the like, is not only desirable from a philanthropic point of view, but an essential condition of safe industrial development, appears to me to be indisputable. It is by such means alone, so far as I can see, that we can hope to check the constant gravitation of industrial society towards *la misère*, until the general progress of intelligence and morality leads men to grapple with the sources of that tendency. If it is said that the carrying out of such arrangements as those indicated must enhance the cost of production, and thus handicap the producer in the race of competition, I venture, in the first place, to doubt the fact; but if it be so, it results that industrial society has to face a dilemma, either horn of which threatens impalement.

On the one hand, a population whose labour is sufficiently remunerated may be physically and morally healthy and socially stable, but may fail in industrial competition by reason of the dearness of its produce. On the other hand, a population whose labour is insufficiently remunerated must become physically and morally unhealthy, and socially unstable; and though it may succeed for a while in industrial competition, by reason of the cheapness of its produce, it must in the end fall, through hideous misery and degradation, to utter ruin.

Well, if these are the only possible alternatives, let us for ourselves and our children choose the former, and, if need be, starve like men. But I do not believe that a stable society made up of healthy, vigorous, instructed, and self-ruling people would ever incur serious risk of that fate. They are not likely to be troubled with many competitors of the same character, just yet; and they may be safely trusted to find ways of holding their own.

Assuming that the physical and moral well-being and the stable social order, which are the indispensable conditions of permanent industrial development, are secured, there remains for consideration the means of attaining that knowledge and skill, without which, even then, the battle of competition cannot be successfully fought. Let us consider how we stand. A vast system of elementary education has now been in operation among us for sixteen years, and has reached all but a very small fraction of the population. I do not think that there is any room for doubt that, on the whole, it has worked well, and that its indirect no less than its direct benefits have been immense. But, as might be expected, it exhibits the defects of all our educational systems—fashioned as they were to meet the wants of a bygone condition of society. There is a widespread, and I think well-justified, complaint that it has too much to do with books and too little to do with things. I am as little disposed as any one can well be to narrow early education and to make the primary school a mere annexe of the shop. And it is not so much in the interests of industry as in that of breadth of culture that I echo the common complaint against the bookish and theoretical character of our primary instruction.

If there were no such things as industrial pursuits, a system of education which does nothing for the faculties of observation, which trains neither the eye nor the hand, and is compatible with utter ignorance of the commonest natural truths, might still be reasonably regarded as strangely imperfect. And when we consider that the instruction and training which are lacking are exactly those which are of most importance for the great mass of our population, the fault becomes almost a crime, the more that there is no practical difficulty in making good these defects. There really is no reason why drawing should not be universally taught, and it is an admirable training for both eye and hand. Artists are born, not made; but everybody may be taught to draw elevations, plans and sections; and pots and pans are as good, indeed better, models for this purpose than the Apollo Belvedere. The plant is not expensive; and there is this excellent quality about drawing of the kind indicated, that it can be tested almost as easily and severely as arithmetic. Such drawings are either right or wrong, and if they are wrong the pupil can be made to see that they are wrong. From the industrial point of view, drawing has the further merit that there is hardly any trade in which the power of drawing is not of daily and hourly utility. In the next place, no good reason, except the want of capable teachers, can be assigned why elementary notions of science should not be an element in general instruction. In this case, again, no experience or elaborate apparatus is necessary. The commonest thing—a candle, a boy's squirt, a piece of chalk—in the hands of a teacher who knows his business may be made the starting points whence children may be led into the regions of science as far as their capacity permits, with efficient exercise of their observational and reasoning faculties on the road. If object lessons often prove trivial failures, it is not the fault of object lessons, but that of the teacher, who has not found out how much the power of teaching a little depends on knowing a great deal, and that thoroughly; and that he has not made that discovery is not the fault of the teachers, but of the detestable system of training them which is widely prevalent.[1]

As I have said, I do not regard the proposal to add these to the present subjects of universal instruction, as made merely in the interests of industry. Elementary science and drawing are just as needful at Eton (where I am happy to say both are now parts of the regular course) as in the lowest primary school. But their importance in the education of the artisan is enhanced, not merely by the fact that the knowledge and skill thus gained—little as they may amount to—will still be of practical utility to him; but further, because they constitute an introduction to that special training which is commonly called "technical education."

I conceive that our wants in this last direction may be grouped under three heads : (1) Instruction in the principles of those branches of science and of art which are peculiarly applicable to industrial pursuits,

[1] Training in the use of simple tools is no doubt very desirable, on all grounds. From the point of view of "culture," the man whose "fingers are all thumbs" is but a stunted creature. But the practical difficulties in the way of introducing handiwork of this kind into elementary schools appear to me to be considerable.

which may be called preliminary scientific education. (2) Instruction in the special branches of such applied science and art, as technical education proper. (3) Instruction of teachers in both these branches. (4) Capacity-catching machinery.

A great deal has already been done in each of these directions, but much remains to be done. If elementary education is amended in the way that has been suggested, I think that the school-boards will have quite as much on their hands as they are capable of doing well. The influences under which the members of these bodies are elected do not tend to secure fitness for dealing with scientific or technical education; and it is the less necessary to burden them with an uncongenial task as there are other organizations, not only much better fitted to do the work, but already actually doing it.

In the matter of preliminary scientific education, the chief of these is the Science and Art Department which has done more during the last quarter of a century for the teaching of elementary science among the masses of the people than any organization which exists either in this or in any other country. It has become veritably a people's university, so far as physical science is concerned. At the foundation of our old universities they were freely open to the poorest but the poorest must come to them. In the last quarter of a century, the Science and Art Department, by means of its classes spread all over the country and open to all, has conveyed instruction to the poorest. The University Extension movement shows that our older learned corporations have discovered the propriety of following suit.

Technical education, in the strict sense, has become a necessity for two reasons. The old apprenticeship system has broken down, partly by reason of the changed conditions of industrial life, and partly because trades have ceased to be "crafts," the traditional secrets whereof the master handed down to his apprentices. Invention is constantly changing the face of our industries, so that "use and wont," "rule of thumb," and the like, are gradually losing their importance, while that knowledge of principles which alone can deal successfully with changed conditions is becoming more and more valuable. Socially, the "master" of four or five apprentices is disappearing in favour of the "employer" of forty, or four hundred, or four thousand "hands," and the odds and ends of technical knowledge, formerly picked up in a shop, are not, and cannot be, supplied in the factory. The instruction formerly given by the master must therefore be more than replaced by the systematic teaching of the technical school.

Institutions of this kind on varying scales of magnitude and completeness, from the splendid edifice set up by the City and Guilds Institute to the smallest local technical school, to say nothing of classes, such as those in technology instituted by the Society of Arts (subsequently taken over by the City Guilds), have been established in various parts of the country, and the movement in favour of their increase and multiplication is rapidly growing in breadth and intensity. But there is much difference of opinion as to the best way in which the technical instruction, so generally desired, should be given. Two courses appear to be practicable: the one is the establishment of special technical schools with a systematic and lengthened course of instruction demanding the employment of the whole time of the pupils. The other is the setting afoot of technical classes, especially evening classes, comprising a short series of lessons on some special topic, which may be attended by persons already earning wages in some branch of trade or commerce.

There is no doubt that technical schools, on the plan indicated under the first head, are extremely costly; and, so far as the teaching of artizans is concerned, it is very commonly objected to them that, as the learners do not work under trade conditions, they are apt to fall into amateurish habits, which prove of more hindrance than service in the actual business of life. When such schools are attached to factories under the direction of an employer who desires to train up a supply of intelligent workmen of course this objection does not apply; nor can the usefulness of such schools for the training of future employers and for the higher grade of the employed be doubtful; but they are clearly out of the reach of the great mass of the people, who have to earn their bread as soon as possible. We must therefore look to the classes, and especially to evening classes, as the great instrument for the technical education of the artizan. The utility of such classes has now been placed beyond all doubt; the only question which remains is to find the ways and means of extending them.

We are here, as in all other questions of social organization, met by two diametrically opposed views. On the one hand, the methods pursued in foreign countries are held up as our example. The State is exhorted to take the matter in hand, and establish a great system of technical education. On the other hand, many economists of the individualist school exhaust the resources of language in condemning and repudiating, not merely the interference of the general government in such matters, but the application of a farthing of the funds raised by local taxation to these purposes. I entertain a strong conviction that, in this country, at any rate, the State had much better leave purely technical and trade instruction alone. But, although my personal leanings are decidedly towards the individualists, I have arrived at that conclusion on merely practical grounds. In fact, my individualism is rather of a sentimental sort, and I sometimes think I should be stronger in the faith if it were less vehemently advocated.[1] I

[1] In what follows I am only repeating and emphasizing opinions which I expressed seventeen years ago, in an Address to the members of the Midland Institute (republished in *Critiques and Addresses* in 1873). I have seen no reason to modify them, notwithstanding high authority on the other side.

am unable to see that civil society is anything but a corporation established for a moral object—namely, the good of its members—and therefore that it may take such measures as seem fitting for the attainment of that which the general voice decides to be the general good. That the suffrage of the majority is by no means a scientific test of social good and evil is unfortunately too true; but, in practice, it is the only test we can apply, and the refusal to abide by it means anarchy. The purest despotism that ever existed is as much based upon that will of the majority (which is usually submission to the will of a small minority) as the freest republic. Law is the expression of the opinion of the majority; and it is law, and not mere opinion, because the many are strong enough to enforce it.

I am as strongly convinced as the most pronounced individualist can be, that it is desirable that every man should be free to act in every way which does not limit the corresponding freedom of his fellow-man. But I fail to connect that great induction of political science with the practical corollary which is frequently drawn from it : that the State—that is, the people in their corporate capacity—has no business to meddle with anything but the administration of justice and external defence. It appears to me that the amount of freedom which incorporate society may fitly leave to its members is not a fixed quantity, to be determined *a priori* by deduction from the fiction called " natural rights "; but that it must be determined by, and vary with, circumstances. I conceive it to be demonstrable that the higher and the more complex the organization of the social body, the more closely is the life of each member bound up with that of the whole ; and the larger becomes the category of acts which cease to be merely self-regarding, and which interfere with the freedom of others more or less seriously.

If a squatter, living ten miles away from any neighbour, chooses to burn his house down to get rid of vermin, there may be no necessity (in the absence of insurance offices) that the law should interfere with his freedom of action ; his act can hurt nobody but himself. But, if the dweller in a street chooses to do the same thing, the State very properly makes such a proceeding a crime, and punishes it as such. He does meddle with his neighbour's freedom, and that seriously. So it might, perhaps, be a tenable doctrine, that it would be needless, and even tyrannous, to make education compulsory in a sparse agricultural population, living in abundance on the produce of its own soil ; but, in a densely populated manufacturing country, struggling for existence with competitors, every ignorant person tends to become a burden upon, and, so far, an infringer of the liberty of, his fellows, and an obstacle to their success. Under such circumstances an education rate is, in fact, a war tax, levied for purposes of defence.

That State action always has been more or less misdirected, and always will be so, is, I believe, perfectly true. But I am not aware that it is more true of the action of men in their corporate capacity than it is of the doings of individuals. The wisest and most dispassionate man in existence, merely wishing to go from one stile in a field to the opposite, will not walk quite straight—he is always going a little wrong, and always correcting himself; and I can only congratulate the individualist who is able to say that his general course of life has been of a less undulatory character. To abolish State action, because its direction is never more than approximately correct, appears to me to be much the same thing as abolishing the man at the wheel altogether, because do what he will, the ship yaws more or less. "Why should I be robbed of my property to pay for teaching another man's children?" is an individualist question, which is not unfrequently put as if it settled the whole business. Perhaps it does, but I find difficulties in seeing why it should. The parish in which I live makes me pay my share for the paving and lighting of a great many streets that I never pass through; and I might plead that I am robbed to smooth the way and lighten the darkness of other people. But I am afraid the parochial authorities would not let me off on this plea; and I must confess I do not see why they should.

I cannot speak of my own knowledge, but I have every reason to believe that I came into this world a small reddish person, certainly without a gold spoon in my mouth, and in fact with no discernible abstract or concrete "rights" or property of any description. If a foot was not, at once, set upon me as a squalling nuisance, it was either the natural affection of those about me, which I certainly had done nothing to deserve, or the fear of the law which, ages before my birth, was painfully built up by the society into which I intruded, that prevented that catastrophe. If I was nourished, cared for, taught, saved from the vagabondage of a wastrel, I certainly am not aware that I did anything to deserve those advantages. And, if I possess anything now, it strikes me that, though I may have fairly earned my day's wages for my day's work, and may justly call them my property—yet, without that organization of society, created out of the toil and blood of long generations before my time, I should probably have had nothing but a flint axe and an indifferent hut to call my own; and even those would be mine only so long as no stronger savage came my way.

So that if society, having—quite gratuitously—done all these things for me, asks me in turn to do something towards its preservation—even if that something is to contribute to the teaching of other men's children—I really, in spite of all my individualist leanings, feel rather ashamed to say no. And if I were not ashamed, I cannot say that I think that society would be dealing unjustly with me in converting the moral obligation into a legal one. There is a manifest unfairness in letting all the burden be borne by the willing horse.

It does not appear to me, then, that there is any valid objection to taxation for purposes of education; but, in the case of technical schools and classes, I think it is practically expedient that such taxation should be local. Our industrial population accumulates in particular towns and districts; these districts are those which immediately profit by technical education; and it is only in them that we can find the men practically engaged in industries, among whom some may reasonably be expected to be competent judges of that which is wanted, and of the best means of meeting the want.

In my belief, all methods of technical training are at present tentative, and, to be successful, each must be adapted to the special peculiarities of its locality. This is a case in which we want twenty years, not of "strong government," but of cheerful and hopeful blundering; and we may be thankful if we get things straight in that time.

The principle of the Bill introduced, but dropped, by the Government last session, appears to me to be wise, and some of the objections to it I think are due to a misunderstanding. The bill proposed in substance to allow localities to tax themselves for purposes of technical education—on the condition that any scheme for such purpose should be submitted to the Science and Art Department, and declared by that Department to be in accordance with the intention of the Legislature.

A cry was raised that the Bill proposed to throw

technical education into the hands of the Science and Art Department. But, in reality, no power of initiation, nor even of meddling with details, was given to that Department—the sole function of which was to decide whether any plan proposed did or did not come within the limits of "technical education." The necessity for such control, somewhere, is obvious. No legislature, certainly not ours, is likely to grant the power of self-taxation without setting limits to that power in some way ; and it would neither have been practicable to devise a legal definition of technical education, nor commendable to leave the question to the Auditor-General to be fought out in the law courts. The only alternative was to leave the decision to an appropriate State authority. If it is asked, what is the need of such control if the people of the localities are the best judges, the obvious reply is that there are localities and localities, and that while Manchester, or Liverpool, or Birmingham, or Glasgow, might, perhaps, be safely left to do as they thought fit, smaller towns, in which there is less certainty of full discussion by competent people of different ways of thinking, might easily fall a prey to crotcheteers.

Supposing our intermediate science teaching and our technical schools and classes are established, there is yet a third need to be supplied, and that is the want of good teachers. And it is necessary not only to get them, but to keep them when you have got them.

It is impossible to insist too strongly upon the fact, that efficient teachers of science and of technology are not to be made by the processes in vogue at ordinary training colleges. The memory loaded with mere bookwork is not the thing wanted—is, in fact, rather worse than useless—in the teacher of scientific subjects. It is absolutely essential that his mind should be full of knowledge and not of mere learning and that what he knows should have been learned in the laboratory rather than in the library. There are happily already, both in London and in the provinces, various places in which such training is to be had, and the main thing at present is to make it in the first place accessible, and in the next indispensable, to those who undertake the business of teaching. But when the well-trained men are supplied, it must be recollected that the profession of teacher is not a very lucrative or otherwise tempting one, and that it may be advisable to offer special inducements to good men to remain in it. These, however, are questions of detail into which it is unnecessary to enter further.

Last, but not least, comes the question of providing the machinery for enabling those who are by nature specially qualified to undertake the higher branches of industrial work, to reach the position in which they may render that service to the community. If all our educational expenditure did nothing but pick one man of scientific or inventive genius, each year, from amidst the hewers of wood and drawers of water, and give him the chance of making the best of his inborn faculties, it would be a very good investment. If there is one such child among the hundreds of thousands of our annual increase, it would be worth any money to drag him either from the slough of misery or from the hotbed of wealth, and teach him to devote himself to the service of his people. Here, again, we have made a beginning with our scholarships and the like, and need only follow in the tracks already worn.

The programme of industrial development briefly set forth in the preceding pages is not what Kant calls a " Hirngespinnst," a cobweb spun in the brain of a Utopian philosopher. More or less of it has taken bodily shape in many parts of the country, and there are towns of no great size or wealth in the manufacturing districts (Keighley, for example) in which almost the whole of it has, for some time, been carried out so far as the means at the disposal of the energetic and public-spirited men who have taken the matter in hand, permitted. The thing can be done; I have endeavoured to show good grounds for the belief that it must be done, and that speedily, if we wish to hold our own in the war of industry. I doubt not that it will be done, whenever its absolute necessity becomes as apparent to all those who are absorbed in the actual business of industrial life as it is to some of the lookers-on.

[Perhaps it is necessary for me to add that technical education is not here proposed as a panacea for social diseases; but simply as a medicament which will help the patient to pass through an imminent crisis.

An ophthalmic surgeon may recommend an operation for cataract in a man who is going blind, without being supposed to undertake that it will cure him of gout. And I may pursue the metaphor so far as to point out, that the surgeon is justified in pointing out that a diet of pork-chops and burgundy will probably kill his patient; though he may be quite unable to suggest a mode of living which will free him from his constitutional disorder.

Mr. Booth asks me, Why do you not propose some plan of your own? Really, that is no answer to my argument that his treatment will make the patient very much worse.]

January 1891.

IN DARKEST ENGLAND

I

The "Times," December 1st, 1890

SIR,—A short time ago a generous and philanthropic friend wrote to me, placing at my disposal a large sum of money for the furtherance of the vast scheme which the General of the Salvation Army has propounded, if I thought it worthy of support. The responsibility of advising my benevolent correspondent has weighed heavily upon me, but I felt that it would be cowardly, as well as ungracious, to refuse to accept it. I have therefore studied Mr. Booth's book with some care, for the purpose of separating the essential from the accessory features of his project, and I have based my judgment—I am sorry to say an unfavourable one—upon the *data* thus obtained. Before communicating my conclusions to my friend, however, I am desirous to know what there may be to be said in arrest of that judgment; and the matter is of such vast public importance that I trust you will aid me by publishing this letter, notwithstanding its length.

There are one or two points upon which I imagine all thinking men have arrived at the same convictions as those from which Mr. Booth starts. It is certain that there is an immense amount of remediable misery among us; that, in addition to the poverty, disease and degradation, which are the consequences of causes beyond human control, there is a vast, probably a very much larger, quantity of misery which is the result of individual ignorance, or misconduct, and of faulty social arrangements. Further, I think it is not to be doubted that, unless this remediable misery is effectually dealt with, the hordes of vice and pauperism will destroy modern civilization as effectually as uncivilized tribes of another kind destroyed the great social organization which precede ours. Moreover, I think all will agree that no reforms and improvements will go to the root of the evil unless they attack it in its ultimate source—namely, the motives of the individual man. Honest, industrious, and self-restraining men will make a very bad social organization prosper; while vicious, idle, and reckless citizens will bring to ruin the best that ever was, or ever will be, invented.

The leading propositions which are peculiar to Mr. Booth, I take to be these:—

(1) That the only adequate means to such reformation of the individual man is the adoption of that form of somewhat corybantic Christianity of

which the soldiers of the Salvation Army are the militant missionaries. This implies the belief that the excitement of the religious emotions (largely by processes described by their employers as "rousing" and "convivial") is a desirable and trustworthy method of permanently amending the conduct of mankind.

I demur to these propositions. I am of opinion that the testimony of history, no less than the cool observation of that which lies within the personal experience of many of us, is wholly adverse to it.

(2) That the appropriate instrument for the propagation and maintenance of this peculiar sacramental enthusiasm is the Salvation Army—a body of devotees, drilled and disciplined as a military organization, and provided with a numerous hierarchy of officers, every one of whom is pledged to blind and unhesitating obedience to the "General," who frankly tells us that the first condition of the service is "implicit, unquestioning obedience." "A telegram from me will send any of them to the uttermost parts of the earth"; every one "has taken service on the express condition that he or she will obey, without questioning, or gainsaying, the orders from headquarters" (*Darkest England*, p. 243).

This proposition seems to me to be indisputable. History confirms it. Francis of Assisi and Ignatius Loyola made their great experiments on the same principle. Nothing is more certain than that a body of religious enthusiasts (perhaps we may even say fanatics) pledged to blind obedience to their chief, is one of the most efficient instruments for effecting any purpose that the wit of man has yet succeeded in devising. And I can but admire the insight into human nature which has led Mr. Booth to leave his unquestioning and unhesitating instruments unbound by vows. A volunteer slave is worth ten sworn bondsmen.

(3) That the success of the Salvation Army, with its present force of 9416 officers " wholly engaged in the work," its capital of three-quarters of a million, its income of the same amount, its 1375 corps at home, and 1499 in the colonies and foreign countries (Appendix, pp. 3 and 4), is a proof that Divine assistance has been vouchsafed to its efforts.

Here I am not able to agree with the sanguine commander-in-chief of the new model, whose labours in creating it have probably interfered with his acquisition of information respecting the fate of previous enterprises of like kind.

It does not appear to me that his success is in any degree more remarkable than that of Francis of Assisi, or that of Ignatius Loyola, than that of George Fox, or even than that of the Mormons, in our own time. When I observe the discrepancies of the doctrinal foundations from which each of these great movements set out, I find it difficult to suppose that supernatural aid has been given to all of them ; still more, that Mr. Booth's smaller measure of success is evidence that it has been granted to him.

But what became of the Franciscan experiment[1]? If there was one rule rather than another on which the founder laid stress, it was that his army of friars should be absolute mendicants, keeping themselves sternly apart from all worldly entanglements. Yet even before the death of Francis, in 1226, a strong party, headed by Elias of Cortona, the deputy of his own appointment, began to hanker after these very things; and, within thirty years of that time, the Franciscans had become one of the most powerful, wealthy, and worldly corporations in Christendom, with their fingers in every sink of political and social corruption, if so be profit for the order could be fished out of it; their principal interest being to fight their rivals, the Dominicans, and to persecute such of their own brethren as were honest enough to try to carry out their founder's plainest injunctions. We also know what has become of Loyola's experiment. For two centuries, the Jesuits have been the hope of the enemies of the Papacy; whenever it becomes too prosperous, they are sure to bring about a catastrophe by their corrupt use of the political and social influence which their organization and their wealth secure.

With these examples of that which may happen to institutions founded by noble men, with high aims, in the hands of successors of a different stamp, armed with despotic authority, before me, common prudence surely requires that before ad-

[1] See note pp. 59, 60.

vising the handing over of a large sum of money to the general of a new order of mendicants I should ask what guarantee there is that, thirty years hence, the "General" who then autocratically controls the action, say, of 100,000 officers pledged to blind obedience, distributed through the whole length and breadth of the poorer classes, and each with his finger on the trigger of a mine charged with discontent and religious fanaticism; with the absolute control, say, of eight or ten millions sterling of capital and as many of income; with barracks in every town, with estates scattered over the country, and with settlements in the colonies—will exercise his enormous powers, not merely honestly, but wisely? What shadow of security is there that the person who wields this uncontrolled authority over many thousands of men shall use it solely for those philanthropic and religious objects which, I do not doubt, are alone in the mind of Mr. Booth? Who is to say that the Salvation Army, in the year 1920, shall not be a replica of what the Franciscan order had become in the year 1260?

The personal character and the intentions of the founders of such organizations as we are considering count for very little in the formation of a forecast of their future; and if they did, it is no disrespect to Mr. Booth to say that he is not the peer of Francis of Assisi. But if Francis's judgment of men was so imperfect as to permit him to appoint an ambitious intriguer of the stamp of Brother Elias his deputy, we have no right to be sanguine about the perspicacity of Mr. Booth in a like matter.

Adding to all these considerations the fact that Mr. Llewelyn Davies, the warmth of whose philanthropy is beyond question, and in whose competency and fairness I, for one, place implicit reliance, flatly denies the boasted success of the Salvation Army in its professed mission, I have arrived at the conclusion that, as at present advised, I cannot be the instrument of carrying out my friend's proposal.

Mr. Booth has pithily characterized certain benevolent schemes as doing sixpennyworth of good and a shilling's worth of harm. I grieve to say that, in my opinion, the definition exactly fits his own project. Few social evils are of greater magnitude than uninstructed and unchastened religious fanaticism; no personal habit more surely degrades the conscience and the intellect than blind and unhesitating obedience to unlimited authority. Undoubtedly, harlotry and intemperance are sore evils, and starvation is hard to bear or even to know of; but the prostitution of the mind, the soddening of the conscience, the dwarfing of manhood are worse calamities. It is a greater evil to have the intellect of a nation put down by organized fanaticism, to see its political and industrial affairs at the mercy of a despot whose chief thought is to make that fanaticism prevail, to watch the degradation of men, who should feel themselves individually responsible for their own and their country's fates, to mere brute instruments ready to the hand of a master for any use to which he may put them.

But that is the end to which, in my opinion, all such organizations as that to which kindly people, who do not look to the consequences of their acts, are now giving their thousands, inevitably tend. Unless clear proof that I am wrong is furnished, another thousand shall not be added by my instrumentality.

I am, Sir, your obedient servant,

T. H. HUXLEY.

NOTE.

An authoritative contemporary historian, Matthew Paris, writes thus of the Minorite, or Franciscan, Friars in England in 1235, just nine years after the death of Francis of Assisi :—

"At this time some of the Minorite brethren, as well as some of the Order of Preachers, unmindful of their profession and the restrictions of their order, impudently entered the territories of some noble monasteries, under pretence of fulfilling their duties of preaching, as if intending to depart after preaching the next day. Under pretence of sickness, or on some other pretext, however, they remained, and, constructing an altar of wood, they placed on it a consecrated stone altar, which they had brought with them, and clandestinely and in a low voice performed mass, and even received the confessions of many of the parishioners, to the prejudice of the priests. . . . And if by chance they were not satisfied with this, they broke forth in insults and threats, reviling every other order except their own, and asserting that all the rest were doomed to damnation, and that they would not spare the soles of their feet till they had exhausted the wealth of their opposers, however great it might be. The religious men, therefore, gave way to them in many points, yielding to avoid scandal, and offending those in power. For they were the councillors and messengers of the nobles, and even secretaries of the Pope, and therefore obtained much secular favour. Some, however, finding themselves opposed at the Court of Rome, were restrained by obvious reasons, and went away in confusion ; for the Supreme Pontiff, with a scowling look, said to them, 'What means this, my brethren? To what lengths are you going? Have you not professed voluntary poverty, and that you would traverse towns and castles and distant places, as the case required, barefooted and unostentatiously in order to preach the word of God in all humility? And do you now presume to usurp these estates to yourselves against the will of the lords of these fees? Your religion appears to be in a great measure dying away, and your doctrines to be confuted.'"

Under date of 1243, Matthew writes :—

"For three or four hundred years or more the monastic order did not hasten to destruction so quickly as their order [Minorites and Preachers] of whom now the brothers, twenty-four years having scarcely elapsed, had first built in England dwellings which rivalled regal palaces in height. These are they who daily expose to view their inestimable treasures, in enlarging their sumptuous edifices, and erecting lofty walls, thereby impudently transgressing the limits of their original poverty and violating the basis of their religion, according to the prophecy of German Hildegarde. When noblemen and rich men are at the point of death, whom they know to be possessed of great riches, they, in their love of gain, diligently urge them, to the injury and loss of the ordinary pastors, and extort confessions and hidden wills, lauding themselves and their own order only, and placing themselves before all others. So no faithful man now believes he can be saved, except he is directed by the counsels of the Preachers and Minorites."—MATTHEW PARIS'S *English History*. Translated by the Rev. J. A. GILES, 1889, Vol. I.

II

The " Times," December 9th, 1890

SIR,—The purpose of my previous letter about Mr. Booth's scheme was to arouse the contributors to the military chest of the Salvation Army to a clear sense of what they are doing. I thought it desirable that they should be distinctly aware that they are setting up and endowing a sect, in many ways analogous to the "Ranters" and "Revivalists" of undesirable notoriety in former times; but with this immensely important difference, that it possesses a strong, far-reaching, centralized organization, the disposal of the physical, moral, and financial strength of which rests with an irresponsible chief, who, according to his own account, is assured of the blind obedience of nearly 10,000 subordinates. I wish them to ask themselves, Ought prudent men and good citizens to aid in the establishment of an organization which, under sundry, by no means improbable, contingencies, may easily become a worse and more dangerous nuisance than the mendicant friars of the middle ages? If this is an academic question, I really do not know what questions deserve to be called practical. As you divined, I purposely omitted any consideration of the details of the Salvationist scheme, and of the principles which animate those who work it, because I desired that the public appreciation of the evils, necessarily inherent in all such plans of despotic social and religious regimentation, should not be obscured by the raising of points of less comparative, however great absolute, importance.

But it is now time to undertake a more particular criticism of *Darkest England.* At the outset of my examination of that work, I was startled to find that Mr. Booth had put forward his scheme with an almost incredibly imperfect knowledge of what had been done and is doing in the same direction. A simple reader might well imagine that the author of *Darkest England* posed as the Columbus, or at any rate the Cortez, of that region. "Go to Mudie's," he tells us, and you will be surprised to see how few books there are upon the social problem. That may or may not be correct; but if Mr. Booth had gone to a certain reading-room not far from Mudie's, I undertake to say that the well-informed and obliging staff of the national library in Bloomsbury would have provided him with more books on this topic, in almost all European languages, than he would read in three months. Has socialism no literature? And what is socialism but an incarnation of the social question? Moreover, I am persuaded that even "Mudie's" resources could have furnished Mr. Booth with the *Life of Lord Shaftesbury* and Carlyle's works. Mr Booth seems to have undertaken to instruct the world without having heard of *Past and Present* or of *Latter-Day Pamphlets;* though, somewhat late in the day, a judicious friend called his attention to them. To those of my contemporaries on whom, as on myself, Carlyle's writings on this topic made an ineffaceable impression forty years ago, who know that, for all that time, hundreds of able and devoted men, both clerical and lay, have worked heart and soul for the permanent amendment of the condition of the poor, Mr. Booth's "Go to Mudie's" affords an apt measure of the depth of his preliminary studies. However, I am bound to admit that these earlier labourers in the field laboured in such a different fashion, that the originality of the plan started by Mr. Booth remains largely unaffected. For them no drums have beat, no trombones brayed; no sanctified buffoonery, after the model of the oration of the Friar in Wallenstein's camp, dear to the readers of Schiller, has tickled the ears of the groundlings on their behalf. Sadly behind the great age of rowdy self-advertisement in which their lot has fallen, they seem not to have advanced one whit beyond John the Baptist and the Apostles, 1800 years ago, in their notions of the way in which the *metanoia*, the change of mind of the ill-doer, is to be brought about. Yet the new model was there, ready for the imitation of those ancient savers of souls. The ranting and roaring mystagogues of some of the most venerable of Greek and Syrian cults also had their processions and banners, their fifes and cymbals and holy chants, their hierarchy of officers to whom the art of making collections was not wholly unknown, and who, as freely as their modern imitators, promised an Elysian future to contributory converts. The success of these antique Salvation armies was enormous. Simon Magus was quite as notorious a personage, and probably had as strong a following, as Mr. Booth. Yet the Apostles, with their old-fashioned ways, would not accept such success as a satisfactory sign of the Divine sanction, nor depart from their own methods of leading the way to the higher life.

I deem it unessential to verify Mr. Booth's statistics. The exact strength of the population of the realm of misery, be it one, two, or three millions, has nothing to do with the efficacy of any means proposed for the highly desirable end of reducing it to a *minimum*. The sole question for consideration at present is whether the scheme, keeping specially in view the spirit in which it is to be worked, is likely to do more good than harm.

Mr. Booth tells us with commendable frankness, that "it is primarily and mainly for the sake of saving the soul that I seek the salvation of the body" (p. 45), which language, being interpreted, means that the propagation of the special Salvationist creed comes first, and the promotion of the physical, intellectual, and purely moral welfare of mankind second in his estimation. Men are to be made sober and industrious, mainly, that, as washed, shorn, and docile sheep, they may be driven into the narrow theological fold which Mr. Booth patronizes. If they refuse to enter, for all their moral cleanliness, they will have to take their place among the goats as sinners, only less dirty than the rest.

I have been in the habit of thinking (and I believe the opinion is largely shared by reasonable men) that self-respect and thrift are the rungs of the ladder by which men may most surely climb out of the slough of despond of want; and I have regarded them as perhaps the most eminent of the practical virtues. That is not Mr. Booth's opinion. For him they are mere varnished sins—nothing better than "Pride re-baptized" (p. 46). Shutting his eyes to the necessary consequences of the struggle for life, the existence of which he accepts as fully as any Darwinian,[1] Mr. Booth tells men, whose evil case is one of those consequences, that envy is a corner-stone of our competitive system. With thrift and self-respect denounced as sin, with the suffering of starving men referred to the sins of the capitalist, the gospel according to Mr. Booth may save souls, but it will hardly save society.

In estimating the social and political influence which the Salvation Army is likely to exert, it is important to reflect that the officers (pledged to blind obedience to their General) are not to confine themselves to the functions of mere deacons and catechists (though, under a General like Cyril, Alexandria knew

[1] See p. 100.

to her cost what even they could effect); they are to
be "tribunes of the people," who are to act as their
gratuitous legal advisers; and, when law is not
sufficiently effective, the whole force of the army is
to obtain what the said tribunes may conceive to be
justice, by the practice of ruthless intimidation.
Society, says Mr. Booth, needs "mothering"; and
he sets forth, with much complacency, a variety of
"cases," by which we may estimate the sort of
"mothering" to be expected at his parental hands.
Those who study the materials thus set before them
will, I think, be driven to the conclusion that the
"mother" has already proved herself a most un-
scrupulous meddler, even if she has not fallen within
reach of the arm of the law.

Consider this "case." A, asserting herself to have
been seduced twice, "applied to our people. We
hunted up the man, followed him to the country,
threatened him with public exposure, and forced from
him the payment to his victim of £60 down, an
allowance of £1 a week, and an insurance policy on
his life for £450 in her favour" (p. 222).

Jedburgh justice this. "We" constitute ourselves
prosecutor, judge, jury, sheriff's officer, all in one;
"we" practise intimidation as deftly as if we were
a branch of another League; and, under threat of
exposure, "we" extort a tolerably heavy hush-money
in payment of our silence.

Well, really my poor moral sense is unable to distinguish these remarkable proceedings of the new

popular tribunate from what in French is called *chantage* and in plain English blackmailing. And when we consider that anybody, for any reason of jealousy, or personal spite, or party hatred, might be thus "hunted," "followed," "threatened," and financially squeezed or ruined, without a particle of legal investigation, at the will of a man whom the familiar charged with the inquisitorial business dare not hesitate to obey, surely it is not unreasonable to ask how far does the Salvation Army, in its "tribune of the people" aspect, differ from a Sicilian Mafia? I am no apologist of men guilty of the acts charged against the person who yet, I think, might be as fairly called a "victim" in this case as his partner in wrongdoing. It is possible that, in so peculiar a case, Solomon himself might have been puzzled to apportion the relative moral delinquency of the parties. However that may be, the man was morally and legally bound to support his child, and any one would have been justified in helping the woman to her legal rights, and the man to the legal consequences (in which exposure is included) of his fault.

The action of the General of the Salvation Army in extorting the heavy fine he chose to impose as the price of his silence, however excellent his motives, appears to me to be as immoral as, I hope, it is illegal.

So much for the Salvation Army as a teacher of questionable ethics and of eccentric economics, as the legal adviser who recommends and practises the extraction of money by intimidation, as the fairy godmother who proposes to "mother" society in a fashion which is not to my taste, however much it may commend itself to some of Mr. Booth's supporters.

  I am, Sir, your obedient servant,

T. H. HUXLEY.

III

The " Times," December 11*th,* 1890

SIR,—When I first addressed you on the subject of the projected operations of the Salvation Army, all that I knew about that body was derived from the study of Mr. Booth's book, from common repute, and from occasional attention to the sayings and doings of his noisy squadrons, with which my walks about London, in past years, have made me familiar. I was quite unaware of the existence of evidence respecting the present administration of the Salvation forces which would have enabled me to act upon the sagacious maxim of the American humourist, "Don't prophesy unless you know." The letter you were good enough to publish has brought upon me a swarm of letters and pamphlets. Some favour me with abuse ; some thoughtful correspondents warmly agree with me, and then proceed to point out how much worthier certain schemes of their own are of my friend's support ; some send valuable encouragement, for which I offer my hearty thanks, and ask them to excuse any more special acknowledgment. But that which I find most to the purpose just now is the revelation made by some of the documents which have reached me, of a fact of which I was wholly ignorant—namely, that persons who have faithfully and zealously served in the Salvation Army, who express unchanged attachment to its original principles and practice, and who have been in close official relations with the "General," have publicly declared that the process of degradation of the organization into a mere engine of fanatical intolerance and personal ambition, which I declared was inevitable, has already set in and is making rapid progress.

It is out of the question, Sir, that I should occupy the columns of the *Times* with a detailed exposition and criticism of these *pièces justificatives* of my forecast. I say criticism, because the assertions of persons who have quitted any society must, in fairness, be taken with the caution that is required in the case of all *ex parte* statements of hostile witnesses. But it is, at any rate, a notable fact that there are parts of my first letter, indicating the inherent and necessary evil consequences of any such organization, which might serve for abstracts of portions of this evidence, long since printed and published under the public responsibility of the witnesses.

Let us ask the attention of your readers, in the first place, to "An ex-Captain's Experience of the Salvation Army," by J. J. R. Redstone, the genuineness of which is guaranteed by the preface (dated April 5th, 1888) which the Rev. Dr. Cunningham

Geikie has supplied. Mr. Redstone's story is well worth reading on its own account. Told in simple, direct, language, such as John Bunyan might have used, it permits no doubt of the single-minded sincerity of the man, who gave up everything to become an officer of the Salvation Army, but, exhibiting a sad want of that capacity for unhesitating and blind obedience on which Mr. Booth lays so much stress, was thrown aside, penniless—no, I am wrong, with 2s. 4d. for his last week's salary—to shift, with his equally devoted wife, as he best might. I wish I could induce intending contributors to Mr. Booth's army chest to read Mr. Redstone's story. I would particularly ask them to contrast the pure simplicity of his plain tale with the artificial pietism and slobbering unction of the letters which Mr. Ballington Booth addresses to his " dear boy " (a married man apparently older than himself), so long as the said " dear boy " is facing brickbats and starvation as per order.

I confess that my opinion of the chiefs of the Salvation Army has been so distinctly modified by the perusal of this pamphlet that I am glad to be relieved from the necessity of expressing it. It will be much better that I should cite a few sentences from the preface written by Dr. Cunningham Geikie, who expresses warm admiration for the early and uncorrupted work of the Salvation Army, and cannot possibly be accused of prejudice against it on religious grounds :—

(1) "The Salvation Army 'is emphatically a family concern. Mr. Booth, senior, is General; one son is chief of the staff, and the remaining sons and daughters engross the other chief positions. It is Booth all over; indeed, like the sun in your eyes, you can see nothing else wherever you turn.' And, as Dr. Geikie shrewdly remarks, 'to be the head of a widely-spread sect carries with it many advantages —not all exclusively spiritual.'

(2) "Whoever becomes a Salvation officer is henceforth a slave, helplessly exposed to the caprice of his superiors."

"Mr. Redstone bore an excellent character both before he entered the army and when he left it. To join it, though a married man, he gave up a situation which he had held for five years and he served Mr. Booth two years, working hard in most difficult posts. His one fault, Major Lawley tells us, was, that he was 'too straight'—that is, too honest, truthful, and manly—or, in other words, too real a Christian. Yet without trial, without formulated charges, on the strength of secret complaints which were never, apparently, tested, he was dismissed with less courtesy than most people would show a beggar—with 2s. 4d. for his last week's salary. If there be any mistake in this matter I shall be glad to learn it."

(3) Dr. Geikie confirms, on the ground of information given confidentially by other officers, Mr. Redstone's assertion that they are watched and reported by spies from headquarters.

(4) Mr. Booth refuses to guarantee his officers any fixed amount of salary. While he and his family of high officials live in comfort, if not in luxury, the pledged slaves whose devotion is the foundation of any true success the Army has met with, often have " hardly food enough to sustain life. One good fellow frankly told me that when he had nothing he just went and begged."

At this point it is proper that I should interpose an apology for having hastily spoken of such men as Francis of Assisi, even for purposes of warning, in connection with Mr. Booth. Whatever may be thought of the wisdom of the plans of the founders of the great monastic orders of the middle ages, they took their full share of suffering and privation, and never shirked in their own persons the sacrifice they imposed on their followers.

I have already expressed the opinion, that whatever the ostensible purpose of the scheme under discussion, one of its consequences will be the setting up and endowment of a new Ranter-Socialist sect. I may now add that another effect will be—indeed, has been—to set up and endow the Booth dynasty with unlimited control of the physical, moral, and financial resources of the sect. Mr. Booth is already a printer and publisher, who, it is plainly declared, utilizes the officers of the Army as agents for advertising and selling his publications; and some of them are so strongly impressed with the belief that active pushing of Mr. Booth's business is the best road to their master's favour, that when the public obstinately refuse to purchase his papers, they buy them themselves and send the proceeds to headquarters. Mr. Booth is also a retail trader on a large scale, and the Dean of Wells has, most seasonably, drawn attention to the very notable banking project which he is trying to float. Any one who follows Dean Plumptre's clear exposition of the principles of this financial operation can have little doubt that, whether they are or are not adequate to the attainment of the first and second of Mr. Booth's ostensible objects, they may be trusted to effect a wide extension of any kingdom in which worldly possessions are of no value. We are, in fact, in sight of a financial catastrophe like that of Law a century ago. Only it is the poor who will suffer.

I have already occupied too much of your space, and yet I have drawn upon only one of the sources of information about the inner working of the Salvation Army at my disposition. Far graver charges than any here dealt with are publicly brought in the others.

I am, Sir, your obedient servant,

T. H. HUXLEY.

P.S.—I have just read Mr. Buchanan's letter in the *Times* of to-day. Mr. Buchanan is, I believe, an imaginative writer. I am not acquainted with his works, but nothing in the way of fiction he has yet achieved can well surpass his account of my opinions and of the purport of my writings.

IV

The " Times," December 20*th*, 1890

SIR,—In discussing Mr. Booth's projects I have hitherto left in the background a distinction which must be kept well in sight by those who wish to form a fair judgment of the influence, for good or evil, of the Salvation Army. Salvationism, the work of "saving souls" by revivalist methods, is one thing; Boothism, the utilization of the workers for the furtherance of Mr. Booth's peculiar projects, is another. Mr. Booth has captured and harnessed with sharp bits and effectual blinkers, a multitude of ultra-Evangelical missionaries of the revivalist school who were wandering at large. It is this skilfully, if somewhat mercilessly, driven team which has dragged the "General's" coach-load of projects into their present position.

Looking, then, at the host of Salvationists proper, from the "captains" downwards (to whom, in my judgment, the family hierarchy stands in the relation of the Old Man of the Sea to Sinbad), as an independent entity, I desire to say that the evidence before me, whether hostile or friendly to the General and his schemes, is distinctly favourable to them. It exhibits them as, in the main, poor, uninstructed, not unfrequently fanatical, enthusiasts, the purity of whose lives, the sincerity of whose belief, and the cheerfulness of whose endurance of privation and rough usage, in what they consider a just cause, command sincere respect. For my part, though I conceive the corybantic method of soul-saving to be full of dangers, and though the theological speculations of these good people are to me wholly unacceptable, yet I believe that the evils which must follow in the track of such errors, as of all other errors, will be largely outweighed by the moral and social improvement of the people whom they convert. I would no more raise my voice against them (so long as they abstain from annoying their neighbours) than I would quarrel with a man, vigorously sweeping out a stye, on account of the shape of his broom, or because he made a great noise over his work. I have always had a strong faith in the principle of the injunction, " Thou shalt not muzzle the ox that treadeth out the corn." If a kingdom is worth a Mass, as a great ruler said, surely the reign of clean living, industry, and thrift is worth any quantity of tambourines and eccentric doctrinal hypotheses. All that I have hitherto said, and propose further to say, is directed against Mr. Booth's extremely clever, audacious, and hitherto successful, attempt to utilize the credit won by all this honest devotion and self-sacrifice for the purposes of his socialistic autocracy.

I now propose to bring forward a little more

evidence as to how things really stand where Mr. Booth's system has had a fair trial. I obtain it, mainly, from a curious pamphlet, the title of which runs:—*The New Papacy. Behind the Scenes in the Salvation Army*, by an ex-Staff Officer. " Make not my Father's house a house of merchandise (John ii. 16)." 1889. Published at Toronto, by A. Britnell. On the cover it is stated that " This is the book which was burned by the authorities of the Salvation Army." I remind the reader, once more, that the statements which I shall cite must be regarded as *ex parte*; all I can vouch for is that, on grounds of internal evidence and from other concurrent testimony respecting the ways of the Booth hierarchy, I feel justified in using them.

This is the picture the writer draws of the army in the early days of its invasion of the Dominion of Canada :—

"Then, it will be remembered, it professed to be the humble handmaid of the existing churches ; its professed object was the evangelization of the masses. It repudiated the idea of building up a separate religious body, and it denounced the practice of gathering together wealth and the accumulation of property. Men and women other than its own converts gathered around it and threw themselves heart and soul into the work, for the simple reason that it offered, as they supposed, a more extended and widely open field for evangelical effort. Ministers everywhere were invited and welcomed to its platforms, majors and colonels were few and far between, and the supremacy and power of the General were things unknown. . . . Care was taken to avoid anything like proselytism ; its converts were never coerced into joining its ranks. . . . In a word, the organization occupied the position of an auxiliary mission and recruiting agency for the various religious bodies. . . . The meetings were crowded, people professed conversion by the score, the public liberally supplied the means to carry on the work in their respective communities ; therefore every corps was wholly self-supporting, its officers were properly, if not luxuriously, cared for, the local expenditure was amply provided, and under the supervision of the secretary, a local member, and the officer in charge, the funds were disbursed in the towns where they were collected, and the spirit of satisfaction and confidence was mutual all around" (pp. 4, 5).

Such was the army as the green tree. Now for the dry :—

" Those who have been daily conversant with the army's machinery are well aware how entirely and radically the whole system has changed, and how, from a band of devoted and disinterested workers, united in the bonds of zeal and charity for the good of their fellows, it has developed into a colossal and aggressive agency for the building up of a system and a sect, bound by rules and regulations altogether subversive of religious liberty and antagonistic to every (other ?) branch of Christian endeavour, and bound hand and foot to the will of one supreme head and ruler. . . . As the work has spread through the country, and as the area of its endeavours has enlarged, each leading position has been filled, one after the other, by individuals strangers to the country, totally ignorant of the sentiments and idiosyncracies of the Canadian people, trained in one school under the teachings and dominance of a member of the Booth family, and out of whom every idea has been crushed, except that of unquestioning obedience to the General and the absolute necessity of going forward to his bidding without hesitation or question (p. 6).

"What is the result of all this? In the first place, whilst material prosperity has undoubtedly been attained, spirituality has been quenched, and, as an evangelical agency, the army has become almost a dead letter. . . . In seventy-five per cent. of its stations its officers suffer need and privation, chiefly on account of the heavy taxation that is placed upon them to maintain an imposing headquarters and a large ornamental staff. The whole financial arrangements are carried on by a system of inflation and a hand-to-mouth extravagance and blindness as to future contingencies. Nearly all of its original workers and members have disappeared (p. 7). . . . In reference to the religious bodies at large the army has become entirely antagonistic. Soldiers are forbidden by its rules to attend other places of worship without the permission of their officers. . . . Officers or soldiers who may conscientiously leave the service or the ranks are looked upon and often denounced publicly as backsliders. . . . Means of the most despicable description have been resorted to in order to starve them back to the service (p. 8). . . . In its inner workings the army system is identical with Jesuitism. . . . That 'the end justifies the means,' if not openly taught, is as tacitly agreed as in that celebrated order" (p. 9).

Surely a bitter, overcharged, anonymous libel is the reflection which will occur to many who read these passages, especially the last. Well, I turn to other evidence which, at any rate, is not anonymous. It is contained in a pamphlet entitled *General Booth, "the Family," and the Salvation Army, showing its Rise, Progress, and Moral and Spiritual Decline*, by S. H. Hodges, LL.B., late Major in the Army, and formerly private secretary to General Booth (Manchester, 1890). I recommend potential contributors to Mr. Booth's wealth to study this little work also. I have learned a great deal from it. Among other interesting novelties, it tells me that Mr. Booth has discovered "the necessity of a third step or blessing, in the work of Salvation. He said to me one day 'Hodges, you have only two barrels to your gun; I have three'" (p. 31). And if Mr. Hodges's description of this third barrel is correct—"giving up your conscience" and "for God and the army, stooping to do things which even honourable worldly men would not consent to do" (p. 32)—it is surely calculated to bring down a good many things, the first principles of morality among them.

Mr. Hodges gives some remarkable examples of the army practice with the "General's" new rifle. But I must refer the curious to his instructive pamphlet. The position I am about to take up is a serious one; and I prefer to fortify it by the help of evidence which, though some of it may be anonymous, cannot be sneered away. And I shall be believed, when I say that nothing but a sense of the great social danger of the spread of Boothism could induce me to revive a scandal, even though it is barely entitled to the benefit of the Statute of Limitations.

On the 7th of July, 1883, you, Sir, did the public a great service by writing a leading article on the notorious "Eagle" case, from which I take the following extract:—

"Mr. Justice Kay refused the application, but he was induced to refuse it by means which, as Mr. Justice Stephen justly remarked, were highly discreditable to Mr. Booth. Mr. Booth filed an affidavit which appears totally to have misled Mr. Justice Kay, as it would have misled any one who regarded it as a frank and honest statement by a professed teacher of religion."

When I addressed my first letter to you I had

never so much as heard of the "Eagle" scandal. But I am thankful that my perception of the inevitable tendency of all religious autocracies towards evil, was clear enough to bring about a provisional condemnation of Mr. Booth's schemes in my mind. Supposing that I had decided the other way, with what sort of feeling should I have faced my friend, when I had to confess that the money had passed into the absolute control of a person, about the character of whose administration this concurrence of damnatory evidence was already extant?

I have nothing to say about Mr. Booth personally, for I know nothing. On that subject, as on several others, I profess myself an agnostic. But, if he is, as he may be, a saint actuated by the purest of motives, he is not the first saint, who, as you have said, has shown himself "in the ardour of prosecuting a well-meant object" to be capable of overlooking "the plain maxims of every day morality." If I were a Salvationist soldier, I should cry with Othello, "Cassio, I love thee; but never more be officer of mine."

I am, Sir, your obedient servant,

T. H. HUXLEY.

V

The " Times," December 24th, 1890

SIR,—If I have any strong points, finance is certainly not one of them. But the financial, or rather fiscal, operations of the General of the Salvation Army, as they are set forth and exemplified in *The New Papacy*, possess that grand simplicity which is the mark of genius; and even I can comprehend them—or, to be more modest, I can portray them in such a manner that every lineament, however harsh, and every shade, however dark, can be verified by published evidence.

Suppose there is a thriving, expanding colonial town; and that, scattered among its artisans and labourers, there is a sprinkling of Methodists, or other such ultra-evangelical good people, doing their best, in a quiet way, to "save souls." Clearly, this is an outpost which it is desirable to capture. "We," therefore, take measures to get up a Salvation "boom" of the ordinary pattern. Enthusiasm is roused. A score or two of soldiers are enlisted into the ranks of the Salvation Army. "We" select the man who promises to serve our purposes best, make a "captain" of him, and put him in command of the "corps." He is very pleased and grateful; and indeed he ought to be. All he has done is that he has given up his trade; that he has promised to work at least nine hours a day in our service (none of your eight-hour nonsense for us) as collector, bookseller, general agent, and anything else we may order him to be. "We," on the other hand, guarantee him nothing whatever; to do so might weaken his faith and substitute worldly for spiritual ties between us. Knowing that, if he exerts himself in a right spirit, his labours will surely be blessed, we content ourselves with telling him that if, after all expenses are paid and our demands are satisfied each week, 25*s*. remains, he may take it. And, if nothing remains, he may take that, and stay his stomach with what the faithful may give him. With a certain grim playfulness, we add that the value of these contributions will be reckoned as so much salary. So long as our "captain" is successful, therefore, a beneficent spring of cash trickles unseen into our treasury; when it begins to dry up we say "God bless you, dear boy," turn him adrift (with or without 2*s*. 4*d*. in his pocket), and put some other willing horse in the shafts.

The "General," I believe, proposes, among other things, to do away with "sweating." May he not as well set a good example by beginning at home?

My little sketch, however, looks so like a monstrous caricature that, after all, I must produce the original from the pages of my Canadian authority. He says that a "captain" "has to pay 10 per cent. of all collections and donations to the divisional fund for the support of his divisional officer, who has also the privilege of arranging for such special meetings as he shall think fit, the proceeds of which he takes away for the general needs of the division. Headquarters, too, has the right to hold such special meetings at the corps and send around such special attractions as its wisdom sees fit, and to take away the proceeds for the purposes it decides upon. . . . He has to pay the rent of his building, either to headquarters or a private individual; he has to send the whole collection of the afternoon meeting of the first Sunday in the month to the 'Extension Fund' at headquarters; he has to pay for the heating, lighting, and cleaning of his hall, together with such necessary repairs as may be needed; he has to provide the food, lodging, and clothing of his cadet, if he has one; headquarters taxes him with so many copies of the army papers each week, for which he has to pay, sold or unsold; and when he has done this, he may take $6 (or $5, being a woman), or such proportion of it as may be left, with which to clothe and feed himself and to pay the rent and provide for the heating and lighting of his quarters. If he has a lieutenant he has to pay him $6 per week, or such proportion of it as he himself gets, and share the house expenses with him. Now, it will be easily understood that at least 60 per cent. of the stations in Canada the officer gets no money at all, and he has to beg specially amongst his people for his house-rent and food. There are few places in the Dominion in which the soldiers do not find their officers in all the food they need; but it must be remembered that the value of the food so received has to be accounted for at headquarters and entered upon the books of the corps as cash received, the amount being deducted from any moneys that the officer is able to take from the week's collections. So that, no matter how much may be specially given, the officer cannot receive more than the value of $6 per week. The officer cannot collect any arrears of salary, as each week has to pay its own expenses; and if there is any surplus cash after all demands are met it must be sent to the 'war chest' at headquarters." *The New Papacy* (pp. 35, 36).

Evidently, Sir, "headquarters" has taken to heart the injunction about casting your bread upon the waters. It casts the crumb of a day or two's work of an emissary and gets back any quantity of loaves of cash, so long as "captains" present themselves to be used up and replaced by new victims. What can be said of these devoted poor fellows except, *O sancta simplicitas!*

But it would be a great mistake to suppose that the money-gathering efficacy of Mr. Booth's fiscal agencies is exhausted by the foregoing enumeration of their regular operations. Consider the following edifying history of the "Rescue Home" in Toronto:—

"It is a fine building in the heart of the city; the

lot cost $7,000, and a building was put up at a cost of $7,000 more, and there is a mortgage on it amounting to half the cost of the whole. The land to-day would probably fetch double its original price, and every year enhances its value . . . . . In the first five months of its existence this institution received from the public an income of $1,812 70c.; out of this $600 was paid to headquarters for rent, $590 52c. was spent upon the building in various ways, and the balance of $622 18c. paid the salaries of the staff and supported the inmates" (pp. 24, 25).

Said I not truly that Mr. Booth's fisc bears the stamp of genius? Who else could have got the public to buy him a "corner lot," put a building upon it, pay all its working expenses: and then, not content with paying him a heavy rent for the use of the handsome present they had made him, they say not a word against his mortgaging it to half its value. And, so far as any one knows, there is nothing to stop headquarters from selling the whole estate to-morrow and using the money as the "General" may direct.

Once more listen to the author of *The New Papacy*, who affirms that "out of the funds given by the Dominion for the evangelization of the people by means of the Salvation Army, one-sixth had been spent in the extension of the Kingdom of God and the other five-sixths had been invested in valuable property, all handed over to Mr. Booth and his heirs and assigns, as we have already stated" (p. 26).

And this brings me to the last point upon which I wish to touch. The answer to all inquiries as to what has become of the enormous personal and real estate which has been given over to Mr. Booth is that it is held ",in trust." The supporters of Mr. Booth may feel justified in taking that statement " on trust." I do not. Anyhow, the more completely satisfactory this "trust" is, the less can any man who asks the public to put blind faith in his integrity and his wisdom object to acquaint them exactly with its provisions. Is the trust drawn up in favour of the Salvation Army ? But what is the legal *status* of the Salvation Army ? Have the soldiers any claim ? Certainly not. Have the officers any legal interest in the "trust"? Surely not. The "General" has taken good care to insist on their renouncing all claims as a condition of their appointment. Thus, to all appearance, the army, as a legal person, is identical with Mr. Booth. And in that case any "trust" ostensibly for the benefit of the army is—what shall we say that is at once accurate and polite ?

I conclude with these plain questions—Will Mr. Booth take counsel's opinion as to whether there is anything in such legal arrangements as he has at present made which prevents him from disposing of the wealth he has accumulated at his own will and pleasure ? Will anybody be in a position to set either the civil or the criminal law in motion against him or his successors if he or they choose to spend every farthing in ways very different from those contemplated by the donors.

I may add that a careful study of the terms of a "Declaration of Trust by William Booth in favour of the Christian Mission," made in 1878, has not enabled persons of much greater competence than myself to answer these questions satisfactorily.

I am, Sir, your obedient servant,

T. H. HUXLEY.

On the 24th December a letter appeared in the *Times* signed " J. S. Trotter," in which the following passages appear :—

" It seems a pity to put a damper on the spirits of those who agree with Professor Huxley in his denunciation of General Booth and all his works. May I give a few particulars as to the ' book ' which was published in Canada ? I had the pleasure of an interview with the author of a book written in Canada. The book was printed at Toronto, and two copies only struck off by the printers ; one of these copies was stolen from the printer, and the quotation sent to you by Professor Huxley was inserted in the book, and is consequently a forgery. The book was published without the consent and against the will of the author.

"So the quotation is not only 'a bitter, overcharged anonymous libel,' as Professor Huxley intimates, but a forgery as well. As to Mr. Hodges, it seems to me to be simply trifling with your readers to bring him in as an authority. He was turned out of the army, out of kindness taken on again, and again dismissed. If this had happened to one of your staff, would his opinion of the *Times* as a newspaper be taken for gospel?"

But in the *Times* of December 29th, Mr. J. S. Trotter writes :—

"I find I was mistaken in saying, in my letter of Wednesday, to the *Times* that Mr. Hodges was dismissed from the service of General Booth, and regret any inconvenience the statement may have caused to Mr. Hodges."

· And, on December 30th the *Times* published a letter from Mr. Hodges in which he says that Mr. Trotter's statements as they regard himself "are the very reverse of truth. I was never turned out of the Salvation Army. Nor, so far as I was made acquainted with General Booth's motives, was I taken on again out of kindness. In order to rejoin the Salvation Army, I resigned the position of manager in a mill where I was in receipt of a salary of £250 per annum, with house rent and one-third of the profits. Instead of this Mr. Booth allowed me £2 per week and house rent."

VI

The " Times," December 26th, 1890

SIR,—I am much obliged to Mr. J. S. Trotter for the letter which you publish this morning. It furnishes evidence which I much desired to possess on the following points :—

1. The author of *The New Papacy* is a responsible, trustworthy person ; otherwise Mr. Trotter would not speak of having had "the pleasure of an interview" with him.

2. After this responsible person had taken the trouble to write a pamphlet of sixty-four closely printed pages, some influence was brought to bear upon him, the effect of which was that he refused his consent to its publication. Mr. Trotter's excellent information will surely enable him to tell us what influence that was.

3. How does Mr. Trotter know that any passage I have quoted is an interpolation ? Does he possess that other copy of the "two" which alone, as he affirms, were printed ?

4. If so, he will be able to say which of the

passages I have cited is genuine and which is not; and whether the tenor of the whole uninterpolated copy differs in any important respect from that of the copy I have quoted.

It will be interesting to hear what Mr. J. S. Trotter has to say upon these points. But the really important thing which he has done is that he has testified, of his own knowledge, that the anonymous author of *The New Papacy* is no mere irresponsible libeller, but a person of whom even an ardent Salvationist has to speak with respect.

I am, Sir, your obedient servant,

T. H. HUXLEY.

[I may add that the unfortunate Mr. Trotter did me the further service of eliciting the letter from Mr. Hodges referred to on p. 91—which sufficiently establishes that gentleman's credit; and leads me to attach full weight to his evidence about the "third barrel."]

January 1891.

VII

The " Times," December 27th, 1890

SIR,—In making use of the only evidence of the actual working of Mr. Booth's autocratic government accessible to me, I was fully aware of the slippery nature of the ground upon which I was treading. For, as I pointed out in my first letter, "no personal habit more surely degrades the conscience and the intellect than blind and unhesitating obedience to unlimited authority." Now we have it, on Mr. Booth's own showing, that every officer of his has undertaken to "obey without questioning or gainsaying the orders from headquarters." And the possible relations of such orders to honour and veracity are demonstrated not only by the judicial deliverance on Mr. Booth's affidavit in the " Eagle" case, which I have already cited; not only by Mr. Bramwell Booth's admission before Mr. Justice Lopes that he had stated what was "not quite correct" because he had "promised Mr. Stead not to divulge" the facts of the case (the *Times*, November 4th, 1885); but by the following passage in Mr. Hodges's account of the reasons of his withdrawal from the Salvation Army :—

"The General and Chief did not and could not deny doing these things; the only question was this, was it right to practise this deception? These points of difference were fully discussed between myself and the Chief of the Staff on my withdrawal, especially the Leamington incident, which was the one that finally drove me to decision. I had come to the conclusion, from the first, that they had acted as they supposed with a single eye to the good of God's cause, and had persuaded myself that the things were, as against the devil, right to be done, that as in battle one party captured and turned the enemy's own guns upon them, so, as they were fighting against the devil, it would be fair to use against him his weapons. And I wrote to this effect to the General" (p. 63).

Now, I do not wish to say anything needlessly harsh, but I ask any prudent man these questions. Could I, under these circumstances, trust any uncorroborated statement emanating from headquarters, or made by the General's order? Had I any reason to doubt the truth of Mr. Hodges's naive confession of the corrupting influence of Mr. Booth's system? And did it not behove me to pick my way carefully through the mass of statements before me, many of them due to people whose moral sense might, by possibility, have been as much blunted by the army discipline in the use of the weapons of the devil as Mr. Hodges affirms that his was?

Therefore, in my third letter, I commenced my illustrations of the practical working of Boothism with the evidence of Mr. Redstone, fortified and supplemented by that of a non-Salvationist, Dr. Cunningham Geikie. That testimony has not been challenged, and until it is, I shall assume that it cannot be. In my fourth letter, I cited a definite statement by Mr. Hodges in evidence of the Jesuitical principles of headquarters. What sort of answer is it to tell us that Mr. Hodges was dismissed the army? A child might expect that some such red herring would be drawn across the trail; and, in anticipation of the stale trick, I added the strong *primâ facie* evidence of the trustworthiness of my witness, in this particular, which is afforded by the "Eagle" case. It was not until I wrote my fourth letter to you, Sir—until the exploitation of the "captains" and the Jesuitry of headquarters could be proved up to the hilt—that I ventured to have recourse to *The New Papacy*. So far as the pamphlet itself goes, this is an anonymous work; and, for sufficient reasons, I did not choose to go beyond what was to be found between its covers. To any one accustomed to deal with the facts of evolution, the Boothism of *The New Papacy* was merely the natural and necessary development of the Boothism of Mr. Redstone's case and of the "Eagle" case. Therefore, I felt fully justified in using it, at the same time carefully warning my readers that it must be taken with due caution.

Mr. Trotter's useful letter admits that such a book was written by a person with whom he had the "pleasure of an interview," and that a version of it (interpolated according to his assertion) was published against the will of the author. Hence I am justified in believing that there is a foundation of truth in certain statements, some of which have long been in my possession, but which for lack of Mr. Trotter's valuable corroboration I have refrained from using. The time is come when I can set forth some of the heads of this information, with the request that Mr. Trotter, who knows all about the business, will be so good as to point out any error that there may be in them. I am bound to suppose that his sole object, like mine, is the elucidation of the truth, and to assume his willingness to help me therein to the best of his ability.

1. "The author of *The New Papacy* is a Mr. Sumner, a person of perfect respectability and greatly esteemed in Toronto, who held a high position in the Army. When he left, a large public meeting, presided over by a popular Methodist minister, passed a vote of sympathy with him."

Is this true or false?

2. "On Saturday last, about noon, Mr. Sumner, the author of the book, and Mr. Fred Perry, the Salvation Army printer, accompanied by a lawyer, went down to Messrs. Imrie and Graham's establish-

ment, and asked for all the manuscript, stereotype plates, &c., of the book. Mr. Sumner explained that the book had been sold to the Army, and, on a cheque for the amount due being given, the printing material was delivered up."

Did these paragraphs appear in the *Toronto Telegram* of April 24th, 1889, or did they not? Are the statements they contain true or false?

3. " Public interest in the fate or probable outcome of that mysterious book called *The New Papacy; or, Behind the Scenes in the Salvation Army*, continues unabated, though the line of proceedings by the publisher and his solicitor, Mr. Smoke, of Watson, Thorne, Smoke, and Masten, has not been altered since yesterday. The book, no doubt, will be issued in some form. So far as known, only one complete copy remains, and the whereabouts of this is a secret which will be profoundly kept. It is safe to say that if the Commissioner kept on guessing until the next anniversary, he would not strike the secluded location of the one volume among five thousand which escaped, when he and his assistant, Mr. Fred Perry, believed they had cast every vestige of the forbidden work into the fiery furnace. On Tuesday last, when the discovery was made that a copy of *The New Papacy* was in existence, Publisher Britnell, of Yonge Street, was at once the suspected holder, and in a short time his book-store was the resort of army

agents sent to reconnoitre" (*Toronto News*, April 28th, 1889).

Is this a forgery, or is it not? Is it in substance true or false?

When Mr. Trotter has answered these inquiries categorically we may proceed to discuss the question of interpolations in Mr. Sumner's book.

I am, Sir, your obedient servant,

T. H. HUXLEY.

[On the 26th of December a letter, signed J. T. Cunningham, late Fellow of University College, Oxford, called forth the following commentary.]

VIII

The " Times," December 29*th*, 1890

SIR,—If Mr. Cunningham doubts the efficacy of the struggle for existence, as a factor in social conditions, he should find fault with Mr. Booth and not with me.

"I am labouring under no delusion as to the possibility of inaugurating the millennium by my social specific. In the struggle of life the weakest will go to the wall and there are so many weak. The fittest in tooth and claw will survive. All that we can do is to soften the lot of the unfit and make their suffering less horrible than it is at present" (*In Darkest England*, p. 44).

That is what Mr. Cunningham would have found if he had read Mr. Booth's book with attention. And, if he will bestow equal pains on my second letter, he will discover that he has interpolated the word ".wilfully" in his statement of my "argument," which runs thus :—" Shutting his eyes to the necessary consequences of the struggle for life, the existence of which he admits as fully as any Darwinian, Mr. Booth tells men whose evil case is one of those consequences that envy is a corner-stone of our competitive system." Mr. Cunningham's physiological studies will have informed him that the process of " shutting the eyes," in the literal sense of the words, is not always wilful; and I propose to illustrate, by the crucial instance his own letter furnishes, that the " shutting of the eyes " of the mind to the obvious consequences of accepted propositions may also be involuntary. At least, I hope so.

1. "Sooner or later," says Mr. Cunningham, "the population problem will block the way once more." What does this mean, except that multiplication, excessive in relation to the contemporaneous means of support, will create a severe competition for those means? And this seems to me to be a pretty accurate " reflection of the conceptions of Malthus " and the other poor benighted folks of a past generation at whom Mr. Cunningham sneers.

2. By way of leaving no doubt upon this subject, Mr. Cunningham further tells us, " The struggle for existence is always going on, of course ; let us thank Darwin for making us realize it." It is pleasant to meet with a little gratitude to Darwin among the *epigoni* who are squabbling over the heritage he conquered for them, but Mr. Cunningham's personal expression of that feeling is hasty. For it is obvious that he has not "realized " the significance of Darwin's teaching—indeed, I fail to discover in Mr. Cunning-

ham's letter any sign that he has even "realized" what he would be at. If the "struggle for existence is always going on;" and if, as I suppose will be granted, industrial competition is one phase of that struggle, I fail to see how my conclusion that it is sheer wickedness to tell ignorant men that "envy" is a corner-stone of competition can be disputed.

Mr. Cunningham has followed the lead of that polished and instructed person, Mr. Ben Tillett, in rebuking me for (as the associates say) attacking Mr. Booth's personal character. Of course, when I was writing, I did not doubt that this very handy, though not too clean, weapon would be used by one or other of Mr. Booth's supporters. And my action was finally decided by the following considerations :—I happen to be a member of one of the largest life insurance societies. There is a vacancy in the directory at present, for which half a dozen gentlemen are candidates. Now, I said to myself, supposing that one of these gentlemen (whose pardon I humbly beg for starting the hypothesis), say Mr. A. in his administrative capacity and as a man of business, has been the subject of such observations as a Judge on the Bench bestowed upon Mr. Booth, is he a person for whom I can properly vote ? And, if I find, when I go to the meeting of the policyholders, that most of them know nothing of this and other evidences of what, by the mildest judgment, must be termed Mr. A.'s unfitness for administrative responsibilities, am I to let them remain in their ignorance ? I leave the answer and its application to men of sense and integrity.

The mention of Mr. Cunningham's ally reminds me that I have omitted to thank Mr. Tillett for his very useful and instructive letter; and I hasten to repair a neglect which I assure Mr. Tillett was more apparent than real. Mr. Tillett's letter is dated December 20th. On the 21st the following pregnant (however unconscious) commentary upon it appeared in *Reynolds's Newspaper*:—

"I have always maintained that the Salvation Army is one of the mightiest Socialistic agencies in the country; and now Professor Huxley comes in to confirm that view. How could it be otherwise? The fantastic religious side of Salvationism will disappear in the course of time, and what will be left? A large number of men and women who have been organized, disciplined, and taught to look for something better than their present condition, and who have become public speakers and not afraid of ridicule. There you have the raw materials for a Socialist army."

Mr. Ben Tillett evidently knows Latin enough to construe *proximus ardet.*

I trust that the public will not allow themselves to be led away by the false issues which are dangled before them. A man really may love his fellow-men; cherish any form of Christianity he pleases; and hold not only that Darwinism is "tottering to its fall," but, if he pleases, the equally sane belief that it never existed ; and yet may feel it his duty to oppose, to the best of his capacity, despotic Socialism in all its forms, and, more particularly, in its Boothian disguise.

I am, Sir, your obedient servant,

T. H. HUXLEY.

[Persons who have not had the advantage of a classical education might fairly complain of my use of the word *epigoni*. To say truth, I had been reading Droysen's *Geschichte des Hellenismus*, and the familiar historical title slipped out unawares. In replying to me, however, the late "Fellow of University College," Oxford, declares he had to look the word out in a Lexicon. I commend the fact to the notice of the combatants over the desirability of retaining the present compulsory modicum of Greek in our Universities.]

IX

The " Times," December 30th, 1890

SIR,—I am much obliged to Messrs. Ranger, Burton, and Matthews for their prompt answer to my questions. I presume it applies to all money collected by the agency of the Salvation Army, though not specifically given for the purposes of the "Christian Mission" named in the deed of 1878; to all sums raised by mortgage upon houses and land so given; and further to funds subscribed for Mr. Booth's various projects, which have no apparent reference to the objects of the "Christian Mission," as defined in the deed. Otherwise, to use a phrase which has become classical, "it does not assist us much." But I must leave these points to persons learned in the law.

And, indeed, with many thanks to you, Sir, for the amount of valuable space which you have allowed me to occupy, I now propose to leave the whole subject. My sole purpose in embarking upon an enterprise, which was extremely distasteful to me, was to prevent the skilful "General," or rather

"Generals," who devised the plan of campaign from sweeping all before them with a rush. I found the pass already held by such stout defenders as Mr. Loch and the Dean of Wells, and, with your powerful help, we have given time for the reinforcements, sure to be sent by the abundant, though somewhat slowly acting, common sense of our countrymen, to come up.

I can no longer be useful, and I return to more congenial occupations.

I am, Sir, your obedient servant,

T. H. HUXLEY.

The following letter appeared in the *Times* of January 2, 1891:—

"DEAR MR. TILLETT,—I have not had patience to read Professor Huxley's letters. The existence of hunger, nakedness, misery, 'death from insufficient food,' even of starvation, is certain, and no agency as yet reaches it. How can any man hinder or discourage the giving of food or help? Why is the house called a workhouse? Because it is for those who cannot work? No, because it was the house to give work or bread. The very name is an argument. I am very sure what Our Lord and His Apostles would do if they were in London. Let us be thankful even to have a will to do the same.

"Yours faithfully,
"HENRY E. CARD. MANNING."

X

The " Times," January 3rd, 1891.

SIR,—In my old favourite, *The Arabian Nights,* the motive of the whole series of delightful narratives is that the sultan, who refuses to attend to reason can be got to listen to a story. May I try whether Cardinal Manning is to be reached in the same way? When I was attending the meeting of the British Association in Belfast nearly forty years ago, I had promised to breakfast with the eminent scholar, Dr. Hincks. Having been up very late the previous night I was behind time ; so hailing an outside car, I said to the driver as I jumped on, " Now drive fast, I am in a hurry." Whereupon he whipped up his horse and set off at a hand-gallop. Nearly jerked off my seat, I shouted, " My good friend, do you know where I want to go ? " " No, yer honner," said the driver, " but, any way, I am driving fast." I have never forgotten this object-lesson in the dangers of ill-regulated enthusiasm. We are all invited to jump on to the Salvation Army car, which Mr. Booth is undoubtedly driving very fast. Some of us have a firm conviction, not only that he is taking a very different direction from that in which we wish to go, but that, before long, car and driver will come to grief. Are we to accept the invitation even at the bidding of the eminent person who appears to think himself entitled to pledge the credit of "Our Lord and His Apostles" in favour of Boothism?

        I am, Sir, your obedient servant,

                        T. H. HUXLEY.

XI

The " Times," January 13*th,* 1891

SIR,—A letter from Mr. Booth-Clibborn, dated January 3rd, appeared in the *Times* of yesterday. This elaborate document occupies three columns of small print—space enough, assuredly, for an effectual reply to the seven letters of mine to which the writer refers, if any such were forthcoming. Mr. Booth-Clibborn signs himself " Commissioner of the Salvation Army for France and Switzerland," but he says that he accepts my "challenge" without the knowledge of his chiefs. Considering the self-damaging character of his letter, it was, perhaps, hardly necessary to make that statement.

Mr. "Commissioner" Booth-Clibborn speaks of my "challenge." I presume that he refers to my request for information about the authorship and fate of *The New Papacy*, in the letter published in the *Times* on December 27th, 1890. The "Commissioner" deals with this matter in paragraph No. 4 of his letter ; and I observe, with no little satisfaction, that he does not venture to controvert any one of the statements of my witnesses. He tacitly admits that the author of *The New Papacy* was a person "greatly esteemed in Toronto," and that he held "a high position in the army"; further, that the Canadian "Commissioner" thought it worth while to pay the printer's bill, in order that the copies already printed off might be destroyed and the pamphlet effectually suppressed. Thus the essential facts of the case are admitted and established beyond question.

How does Mr. Booth-Clibborn try to explain them away?

"Mr. Sumner, who wrote the little book in a hot fit, soon regretted it (as any man would do whose conscience showed him in a calmer moment, when his 'respectability' returned with his repentance, that he had grossly misrepresented), and just before it appeared offered to order its suppression if the army would pay the costs already incurred, and which he was unable to bear."

The New Papacy fills sixty closely-printed duodecimo pages. It is carefully written, and for the most part in studiously moderate language; moreover, it contains many precise details and figures, the ascertainment of which must have taken much time and trouble. Yet, forsooth, it was written in "a hot fit."

I sincerely hope, for the sake of his own credit, that Mr. "Commissioner" Booth-Clibborn does not know as much about this melancholy business as I do. My hands are unfortunately tied, and I am not at liberty to use all the information in my possession. I must content myself with quoting the following passage from the preface to *The New Papacy*:—

"It has not been without considerable thought and a good deal of urging that the following pages have been given to the public. But though we would have shrunk from a labour so distasteful, and have gladly avoided a notoriety anything but pleasant to the feelings, or conducive to our material welfare, we have felt that in the interests of the benevolent public, in the interests of religion, in the interests of a band of devoted men and women whose personal ends are being defeated, and the fruit of whose labour is being destroyed, and, above all, in the interests of that future which lies before the Salvation Army itself if purged and purified in its executive and returned to its original position in the ranks of Canadian Christian effort, it is no more than our duty to throw such light as we are able upon its true inwardness, and with that object and for the furtherance of those ends we offer our pages to the public view."

The preface is dated April, 1889. According to the statement in the *Toronto Telegram*, which Mr. "Commissioner" Booth-Clibborn does not dare to dispute, his Canadian fellow "Commissioner" bought and destroyed the whole edition of *The New Papacy* about the end of the third week in April.

It is clear that the writer of the paragraph quoted from the preface was well out of a "hot fit" if he had ever been in one, while he had not entered on the stage of repentance within three weeks of that time. Mr. "Commissioner" Booth-Clibborn's scandalous insinuations that Mr. Sumner was bribed by "a few sovereigns," and that he was "bought off," in the face of his own admission that Mr. Sumner "offered to order its suppression if the army would pay the costs already incurred, and which he was unable to bear," is a crucial example of that Jesuitry with which the officials of the army have been so frequently charged.

Mr. "Commissioner" Booth-Clibborn says that when "London headquarters heard of the affair, it disapproved of the action of the Commissioner." That circumstance indicates that headquarters is not wholly devoid of intelligence; but it has nothing to do with the value of Mr. Sumner's evidence, which is all I am concerned about. Very likely London headquarters will disapprove of its French "Commissioner's" present action. But what then? The upshot of all this is that Mr. Booth-Clibborn has made as great a blunder as simple Mr. Trotter did. The pair of Balaams greatly desired to curse, but have been compelled to bless. They have, between them, completely justified my reliance on Mr. Sumner as a perfectly trustworthy witness; and neither of them has dared to challenge the accuracy of one solitary statement made by that worthy gentleman, whose full story I hope some day or other to see set

H

before the public. Then the true causes of his action will be made known.

Paragraph 2 of the "Commissioner's" letter says many things, but not much about Mr. Hodges. The columns of the *Times* recently showed that Mr. Hodges was able to compel an apology from Mr. Trotter. I leave it to him to deal with the "Commissioner."

As to the "Eagle" case, treated of in paragraph No. 3, a gentleman well versed in the law, who was in Court during the hearing of the appeal, has assured me that the argument was purely technical; that the facts were very slightly gone into; and that, so far as he knows, no dissenting comment was made on the strictures of the Judge before whom the case first came. Moreover, in the judgment of the Master of the Rolls, fully recorded in the *Times* of February 14th, 1884, the following passages occur:—

"The case had been heard by a learned Judge, who had exercised his discretion upon it, and the Court would not interfere with his discretion unless they could see that he was wrong. The learned Judge had taken a strong view of the conduct of the defendant, but nevertheless had said that he would have given relief if he could have seen how far protection and compensation could be given. And if this Court differed from him in that view, and could give relief without forfeiture, they would be acting on his own principle in doing so. Certain suggestions had been made with that view, and the Court had to consider the case under all the circumstances. . . . He himself (the Master of the Rolls) considered that it was probable the defendant, with his principles, had intended to destroy the property as a public-house, and that it was not right thus to take property under a covenant to keep it up as a public-house, intending to destroy it as such. He did not, however, think this was enough to deprive him of all relief. . . . The defendant could only expect severe terms."

Yet, Sir, Mr. "Commissioner" Booth-Clibborn, this high official of the Salvation Army, has the audacity to tell the public that if I had made inquiries I should have found that "in the Court of Appeal the Judge reversed the decision of his predecessor as regards seven-eighths of the property, and the General was declared to have acted all along with straightforwardness and good faith."

But the nature of Mr. "Commissioner" Booth-Clibborn's conceptions of straightforwardness and good faith is so marvellously illustrated by the portions of his letter with which I have dealt that I doubt not his statements are quite up to the level of the "Army" Regulations and Instructions in regard to those cardinal virtues. As I pointed out must be the case, the slave is subdued to that he works in.

For myself, I must confess that the process of wading through Mr. "Commissioner's" verbose and clumsy pleadings has given me a "hot fit," which,

I undertake to say, will be followed by not so much as a passing shiver of repentance. And it is under the influence of the genial warmth diffused through the frame—on one of those rare occasions when one may be "angry and sin not"—that I infringe my resolution to trouble you with no more letters. On reflection, I am convinced that it is undesirable that the public should be misled, for even a few days, by misrepresentations so serious.

I am copiously abused for speaking of the Jesuitical methods of the superior officials of the Salvation Army. But the following facts have not been, and, I believe, cannot be, denied :—

1. Mr. Booth's conduct in the "Eagle" case has been censured by two of the Judges.

2. Mr. Bramwell Booth admitted before Mr. Justice Lopes that he had made an untrue statement because of a promise he had made to Mr. Stead.[1]

And I have just proved that Mr. "Commissioner" Booth-Clibborn asserts the exact contrary of that which your report of the judgment of the Master of the Rolls tells us that distinguished judge said.

Under these circumstances, I think that my politeness in applying no harder adjective than "Jesuitical" to these proceedings is not properly appreciated.

I am, Sir, your obedient servant,

T. H. HUXLEY.

[1] This statement has been disputed, but not yet publicly. (See p. 121.)

XII

The " Times," January 22nd, 1891

Sir,—I think that your readers will be interested in the accompanying opinion, written in consultation with an eminent Chancery Queen's Counsel, with which I have been favoured. It will be observed that this important legal deliverance justifies much stronger language than any which I have applied to the only security (?) for the proper administration of the funds in Mr. Booth's hands which appears to be in existence.

I am, Sir, your obedient servant,
T. H. Huxley.

1, Dr. Johnson's Buildings, Temple, E.C.,
January 14, 1891.

Mr. Booth's Declaration of Trust Deed, 1878.

" I am of opinion, subject to the question whether there may be any provision in the Charitable Trusts Acts which can be made available for enforcing some scheme for the appropriation of the property, and with regard to the real and leasehold properties whether the conveyances and leases are not altogether void, as frauds on the Mortmain Acts, that nothing can be done to control or to interfere with Booth in the disposition or application of the properties or moneys purported to be affected by the deed.

"As to the properties vested in Booth himself, it appears to me that such are placed absolutely under his power and control both as to the disposal and application thereof, and that there are no trusts for any specific purposes declared which could be enforced, and that there are no defined persons nor classes of persons who can claim to be entitled to the benefit of them, or at whose instance they could be enforced by any legal process.

"As to the properties (if any) vested in trustees appointed by Booth, it appears to me that the only person who has a *locus standi* to enforce these trusts is Booth himself, and that he would have absolute power over the trusts and the property, and might deal with the property as he pleased, and that, as in the former case, nothing could be done in the way of enforcing any trusts against him.

"As to the moneys contributed or raised by mortgage for the general purposes of the mission, it appears to me that Booth may expend them as he pleases, without being subject to any legal control, and that he cannot even be compelled to publish any balance-sheets.

"Whether there are any provisions in the Charitable Trusts Acts which could be made available for enforcing some scheme for the application of the property or funds is a question to which I should require to give a closer consideration should it become necessary to go into it; but at present, after perusing these Acts, and especially 16 and 17 Vict., c. 137, and 18 and 19 Vict., c. 124, I cannot see how they could be made applicable to the trusts as declared in this deed.

"As to the Mortmain Acts, the matter is clearly charitable, and unless in the conveyances and leases to Booth or to the trustees (if any) named by him, all the provisions of the Acts have been complied with, and the deeds have been enrolled under the Acts, they would be void. It is probable, however, that every conveyance and lease has been taken without disclosing any charitable trust, for the purpose of preventing it from being void on the face of it. It is to be noted that the deed is a mere deed poll by Booth himself, without any other party to it, who, as a contracting party, would have a right to enforce it

"Whether there are any objects of the trust I cannot say. If there is, as the recital indicates, a society of enrolled members called 'The Christian Mission,' those members would be objects of the trust, but then, it appears to me, Booth has entire control and determination of the application. And, as to the trusts enuring for the benefit of the 'Salvation Army,' I am not aware what is the constitution of the 'Salvation Army,' but there is no reference whatever to any such body in the deed. I have understood the army as being merely the missionaries, and not the society of worshippers.

"If there is no Christian Mission Society of enrolled members, then there are no objects of the trust. The trusts are purely religious, and trading is entirely beyond its purposes. Booth can 'give away' the property, simply because there is no one who has any right to prevent his doing so.

"ERNEST HATTON."

It is probably my want of legal knowledge which prevents me from appreciating the value of the professed corrections of Mr. Hatton's opinion contained in the letters of Messrs. Ranger, Burton, and Matthew, *Times*, January 28th and 29th, 1891.

The note on page 116 refers to a correspondence, incomplete at the time fixed for the publication of my pamphlet, the nature of which is sufficiently indicated by the subjoined extracts from Mr. Stead's letter in the *Times* of January 20th, and from my reply in the *Times* of January 24th. Referring to the paragraphs numbered 1, 2, at the end of my letter XI., Mr. Stead says:—

"On reading this, I at once wrote to Professor Huxley, stating that, as he had mentioned my name, I was justified in intervening to explain that, so far as the second count in his indictment went—for the Eagle dispute is no concern of mine—he had been misled by an error in the reports of the case which appeared in the daily papers of November 4, 1885. I have his reply to-day, saying that I had better write to you direct. May I ask you, then, seeing that my name has been brought into the affair, to state that, as I was in the dock when Mr. Bramwell Booth was in the witness-box, I am in a position to give the most unqualified denial to the statement as to the alleged admission on his part of falsehood? Nothing was heard in Court of any such admission. Neither the prosecuting counsel nor the Judge who tried the case ever referred to it, although it would obviously have had a direct bearing on the credit of the witness; and the jury, by acquitting Mr. Bramwell Booth, showed that they believed him to be a witness of truth. But fortunately the facts can be verified beyond all gainsaying by a reference to the official shorthand-writer's report of the evidence. During the hearing of the case for the prosecution, Inspector Borner was interrupted by the Judge, who said :—

"'I want to ask you a question. During the whole of that conversation, did Booth in any way suggest that that child had been sold?' Borner replied :—

"'Not at that interview, my Lord.'

"It was to this that Mr. Bramwell Booth referred when, after examination, cross-examination, and re-examination, during which no suggestion had been made that he had ever made the untrue statement now alleged against him, he asked and received leave from the Judge to make the following explanation, which I quote from the official report :—

"'Will you allow me to explain a matter mentioned yesterday in reference to a question asked by your Lordship some days ago with respect to one matter connected with my conduct? Your Lordship asked, I think it was Inspector Borner, whether I had said to him at either of our interviews that the child was sold by her parents, and he replied, "No." That is quite correct; I did not say so to him, and what I wish to say now is that I had been specially requested by Mr. Stead and had given him a promise that I would not under any circumstances divulge the fact of that sale to any person which would make it at all probable that any trouble would be brought upon the persons who had taken part in this investigation.' (Central Criminal Court Reports, Vol. CII., part 612, pp. 1,035-6.)

"In the daily papers of the following day this statement was misreported as follows:—

"'I wish to explain, in regard to your Lordship's condemnation of my having said "No" to Inspector Borner when he asked me whether the child had been sold by her parents—the reason why I stated what was not correct was that I had promised Mr. Stead not to divulge the fact of the sale to any person which would make it probable that any trouble should be brought on persons taking part in this proceeding.'

"Hence the mistake into which Professor Huxley has unwittingly fallen.

"I may add that, so far from the statement never having been challenged for five years, it was denounced as 'a remarkably striking lie' in the *War Cry* of November 14th, and again the same official organ of the Salvation Army of November 18th specifically adduced this misreport as an instance of ' the most disgraceful way' in which the reports of the trial were garbled by some of the papers. What, then, becomes of one of the two main pillars of Professor Huxley's argument?"

In my reply, I point out that, on the 10th of January, Mr. Stead addressed to me a letter, which commences thus:—" I see in the *Times* of this morning that you are about to republish your letters on Booth's book."

I replied to this letter on the 12th of January:—

"DEAR MR. STEAD,—I charge Mr. Bramwell Booth with nothing. I simply quote the *Times* report, the accuracy of which, so far as I know, has never been challenged by Mr Booth. I say I quote the *Times* and not Mr. Hodges,[1] because I took some pains about the verification of Mr. Hodges' citation.

"I should have thought it rather appertained to Mr. Bramwell Booth to contradict a statement which refers, not to what you heard, but to what he said. However, I am the last person to wish to give circulation to a story which may not be quite correct; and I will take care, if you have no objection (your letter is marked 'private'), to make public as much of your letter as relates to the point to which you have called my attention.

"I am, yours very faithfully,

"T. H. HUXLEY."

To this Mr. Stead answered, under date of January 13th, 1891:—

"DEAR PROFESSOR HUXLEY,—I thank you for your letter of the 12th inst. I am quite sure you would not wish to do any injustice in this matter. But, instead of publishing any extract from my letter

[1] This is a slip of the pen. Mr. Hodges had nothing to do with the citation of which I made use.

might I ask you to read the passage as it appears in the verbatim report of the trial which was printed day by day, and used by counsel on both sides, and by the Judge during the case? I had hoped to have got you a copy to-day, but find that I was too late. I shall have it first thing to-morrow morning. You will find that it is quite clear, and conclusively disposes of the alleged admission of untruthfulness. Again thanking you for your courtesy,

"I am, yours faithfully,

"W. T. STEAD."

Thus it appears that the letter which Mr. Stead wrote to me on the 13th of January does not contain one word of that which he says it contains, in the statement which appears in the *Times* to-day. Moreover, the letter of mine to which Mr. Stead refers in his first communication to me, is not the letter which appeared on the 13th, as he states, but that which you published on December 27th, 1890. Therefore, it is not true that Mr. Stead wrote "at once." On the contrary, he allowed nearly a fortnight to elapse before he addressed me on the 10th of January, 1891. Furthermore, Mr. Stead suppresses the fact that, since the 13th of January, he has had in his possession my offer to publish his version of the story; and he leads the reader to suppose that my only answer was that he "had better write to you direct." All the while, Mr. Stead knows perfectly well that I was withheld from making public use of his letter of the 10th by nothing but my scruples about using a document which was marked "private"; and that he did not give me leave to quote his letter of the 10th of January until after he had written that which appeared yesterday.

And I add:

As to the subject-matter of Mr. Stead's letter, the point which he wishes to prove appears to be this—that Mr. Bramwell Booth did not make a false statement, but that he withheld from the officers of justice, pursuing a most serious criminal inquiry, a fact of grave importance, which lay within his own knowledge. And this because he had promised Mr. Stead to keep the fact secret. In short, Mr. Bramwell Booth did not say what was wrong; but he did what was wrong.

I will take care to give every weight to the correction. Most people, I think, will consider that one of the "main pillars of my argument," as Mr. Stead is pleased to call them, has become very much strengthened.

LEGAL OPINIONS RESPECTING "GENERAL" BOOTH'S ACTS.

IN referring to the course of action adopted by "General" Booth and Mr. Bramwell Booth in respect of their legal obligations to other persons, or to the criminal and civil law, I have been as careful as I was bound to be, to put any difficulties suggested by mere lay common-sense in an interrogative or merely doubtful form; and to confine myself, for any positive expressions, to citations from published declarations of the judges before whom the acts of "General" Booth came; from reports of the Law Courts; and from the deliberate opinions of legal experts. I have now some further remarks to make on these topics.

I. The observations at p. 120 express, with due reserve, the impression which the counsel's opinions, quoted by "General" Booth's solicitors, made on my mind. They were written and sent to the printer before I saw the letter from a "Barrister *not* Practising on the Common Law Side," and those from Messrs. Clarke and Calkin and Mr. George Kebbell, which appeared in the *Times* of the 3rd and 4th February.

These letters fully bear out the conclusion which I had

formed, but which it would have been presumptuous on my part to express, that the opinions cited by "General" Booth's solicitors were like the famous broken tea-cups "wisely ranged for show;" and that, as Messrs. Clarke and Calkin say, they "do not at all meet the main points on which Mr. Hatton advised." I do not think that any one who reads attentively the able letter of "A Barrister *not* Practising on the Common Law Side" will arrive at any other conclusion ; or who will not share the very natural desire of Mr. Kebbell to be provided with clear and intelligible answers to the following inquiries :—

(1) Does the trust deed by its operation empower any one legally to call upon Mr. Booth to account for the application of the funds?

(2) In the event of the funds not being properly accounted for, is any one, and, if so, who, in a position to institute civil or criminal proceedings against any one, and whom, in respect of such refusal or neglect to account?

(3) In the event of the proceedings, civil or criminal, failing to obtain restitution of misapplied funds, is or are any other person or persons liable to make good the loss?

On the 24th of December, 1890, a letter of mine appeared in the *Times* (No. V. above) in which I put questions of the same import, and asked Mr. Booth if he would not be so good as to take counsel's opinion on the "trusts" of which so much has been heard and so little seen, not as they stood in 1878, or in 1888, but as they stand now? Six weeks have elapsed and I wait for a reply.

It is true that Dr. Greenwood has been authorized by Mr. Booth to publish what he calls a "Rough outline of the intended Trust Deed" (*General Booth and His Critics*, p. 120), but unfortunately we are especially told that it "*does not profess to be an absolutely accurate analysis.*" Under these circumstances I am afraid that neither lawyers, nor laymen of moderate intelligence, will pay much attention to the assertion, that "*it gives a fair idea of the general effect of the draft,*" even although "*the words in quotation marks are taken from it verbatim.*"

These words, which I give in italics, (1) define the purposes of the scheme to be "*for the social and moral regeneration and improvement of persons needy, destitute, degraded, or criminal, in some manner indicated, implied, or suggested in the book called* ' *In Darkest England.*'" Whence I apprehend that, if the whole funds collected are applied to "mothering society" by the help of speculative attorney "tribunes of the people," the purposes of the trust will be unassailably fulfilled. (2) The name is to be "*Darkest England Scheme,*" (3) the General of the Salvation Army is to be "*Director of the Scheme.*" Truly valuable information all this! But taking it for what it is worth, the public must not be misled into supposing that it has the least bearing upon the questions to which neither I, nor anybody else, has yet been able to obtain an intelligible answer, and that is, where are

the vast funds which have been obtained, in one way
or another, during the last dozen years in the name
of the Salvation Army? Where is the presumably
amended Trust Deed of 1888? I ask once more:
Will Mr. Booth submit to competent and impartial
legal scrutiny the arrangements by which he and his
successors are prevented from dealing with the funds
of the so-called "army-chest" exactly as he or they
may please?

II. With respect to the "Eagle" case, I am advised
that Dr. Greenwood, whose good faith I do not
question, has been misled into misrepresenting it
in the appendix to his pamphlet. And certainly,
the evidence of authoritative records which I
have had the opportunity of perusing, appears to
my non-legal mind to be utterly at variance with the
statement to which Dr. Greenwood stands committed.
I may observe further, that the excuse alleged on
behalf of Mr. Booth, that he signed the affidavit set
before him by his solicitors without duly considering
its contents, is one which I should not like to have
put forward were the case my own. It may be, and
often is, necessary for a person to sign an affidavit
without being able fully to appreciate the technical
language in which it is couched. But his solicitor
will always instruct him as to the effect of these
terms. And, in this particular case, where the whole
matter turns on Mr. Booth's personal intentions, it
was his plainest duty to inquire, very seriously, whether

the legal phraseology employed would convey neither more nor less than such intentions to those who would act on the affidavit, before he put his name to it.

III. With respect to Mr. Bramwell Booth's case I refer the reader to p. 126.

IV. As to Mr. Booth-Clibborn's misrepresentations, see above, p. 114–115.

Thus much for the legal questions which have been raised by various persons since the first edition of the pamphlet was published.

DR. GREENWOOD'S "GENERAL BOOTH AND HIS CRITICS."

So far as I am concerned, there is little or nothing in this *brochure* beyond a reproduction of the vituperative stuff which has been going the round of those newspapers which favour "General" Booth for some weeks. Those who do not want to see the real worth of it all will not read the preceding pages; and those who do will need no help from me.

I fear, however, that in justice to other people, I must put one of Dr. Greenwood's paragraphs in the pillory. He says that I have "built up, on the flimsy foundation of stories told by three or four deserters from the Army,"(p. 114) a sweeping indictment against General Booth. This is the sort of thing to which I am well accustomed at the hands of anonymous newspaper writers. But in view of the following easily verifiable statements, I do not think that an educated, and I have no doubt, highly respectable gentleman like Dr. Greenwood can, in cold blood, contemplate that assertion with satisfaction.

The persons here alluded to as "three or four deserters from the army" are :—

(1) Mr. Redstone, for whose character Dr. Cunningham Geikie is guarantee, and whom it has been left to Dr. Greenwood to attempt to besmirch.

(2) Mr. Sumner, who is a gentleman quite as worthy of respect as Dr. Greenwood, and whose published evidence not one of the champions of the Salvation Army has yet ventured to impugn.

(3) Mr. Hodges, similarly libelled by that unhappy meddler Mr. Trotter, who was compelled to the prompt confession of his error (see p. 91).

(4) Notwithstanding this evidence of Mr. Trotter's claims to attention, Dr. Greenwood quotes a statement of his as evidence that a statement quoted by me from Mr. Sumner's work is a "forgery." But Dr. Greenwood unfortunately forgets to mention that on the 27th December, 1890 (Letter No. VII. above), Mr. Trotter was publicly required to produce proof of his assertion; and that he has not thought fit to produce that proof.

If I were disposed to use to Dr. Greenwood language of the sort he so freely employs to me, I think that he could not complain of a handsome scolding. For what is the real state of the case? Simply this—that having come to the conclusion, from the perusal of *In Darkest England*, that "General" Booth's colossal scheme (as apart from the local action of Salvationists) was bad in principle and must produce certain evil consequences, and having warned the public to that effect, I quite unexpectedly found my hands full of evidence that the exact evils predicted had, in fact, already shown themselves on a great scale; and, carefully warning the public to criticize this evidence, I produced a small part of it. When Dr. Greenwood talks about my want of "regard to the opinion of the nine thousand odd who still remain among the faithful" (p. 114) he commits an imprudence. He would obviously be surprised to learn the extent of the support, encouragement, and information which I have received from active and sincere members of the Salvation Army—but of which I can make no use, because of the terroristic discipline and systematic espionage which my correspondents tell me is enforced by its chief. Some of these days, when nobody can be damaged by their use, a curious light may be thrown upon the inner workings of the organization which we are bidden to regard as a happy family, by these documents.

THE SALVATION ARMY.

ARTICLES OF WAR.

TO BE SIGNED BY ALL WHO WISH TO BE ENTERED ON THE ROLL AS SOLDIERS.

HAVING received with all my heart the Salvation offered to me by the tender mercy of Jehovah, I do here and now publicly acknowledge God to be my Father and King, Jesus Christ to be my Saviour, and the Holy Spirit to be my Guide, Comforter, and Strength ; and that I will, by His help, love, serve, worship, and obey this glorious God through all time and through all eternity.

BELIEVING solemnly that The Salvation Army has been raised up by God, and is sustained and directed by Him, I do here declare my full determination, by God's help, to be a true Soldier of The Army till I die.

> I am thoroughly convinced of the truth of the Army's teaching.
> I believe that repentance towards God, faith in our Lord Jesus Christ, and conversion by the Holy Spirit, are necessary to Salvation, and that all men may be saved.
> I believe that we are saved by grace, through faith in our Lord Jesus Christ, and he that believeth hath the witness of it in himself. I have got it. Thank God !
> I believe that the Scriptures were given by inspiration of God, and that they teach that not only does continuance in the favour of God depend upon continued faith in, and obedience to, Christ, but that it is possible for those who have been truly converted to fall away and be eternally lost.

I believe that it is the privilege of all God's people to be "wholly sanctified," and that "their whole spirit and soul and body" may "be preserved blameless unto the coming of our Lord Jesus Christ." That is to say, I believe that after conversion there remain in the heart of the believer inclinations to evil, or roots of bitterness, which, unless overpowered by Divine grace, produce actual sin; but these evil tendencies can be entirely taken away by the Spirit of God, and the whole heart thus cleansed from anything contrary to the will of God, or entirely sanctified, will then produce the fruit of the Spirit only. And I believe that persons thus entirely sanctified may, by the power of God, be kept unblamable and unreprovable before Him.

I believe in the immortality of the soul; in the resurrection of the body; in the general judgment at the end of the world; in the eternal happiness of the righteous; and in the everlasting punishment of the wicked.

THEREFORE, I do here, and now, and for ever, renounce the world with all its sinful pleasures, companionships, treasures, and objects, and declare my full determination boldly to show myself a Soldier of Jesus Christ in all places and companies, no matter what I may have to suffer, do, or lose, by so doing.

I do here and now declare that I will abstain from the use of all intoxicating liquors, and also from the habitual use of opium, laudanum, morphia, and all other baneful drugs, except when in illness such drugs shall be ordered for me by a doctor.

I do here and now declare that I will abstain from the use of all low or profane language; from the taking of the name of God in vain; and from all impurity, or from taking part in any unclean conversation or the reading of any obscene book or paper at any time, in any company, or in any place.

I do here declare that I will not allow myself in any falsehood, deceit, misrepresentation, or dishonesty; neither will I practise any fraudulent conduct, either in my business, my home, or in any other relation in which I may stand to my fellow men, but that I will deal truthfully, fairly, honourably, and kindly with all those who may employ me or whom I may myself employ.

I do here declare that I will never treat any woman, child, or other person, whose life, comfort, or happiness may be placed within my power, in an oppressive, cruel, or cowardly manner, but that I will protect such from evil and danger so far as I can, and promote, to the utmost of my ability, their present welfare and eternal salvation.

I do here declare that I will spend all the time, strength, money, and influence I can in supporting and carrying on this War, and that I will endeavour to lead my family, friends, neighbours, and all others whom I can influence, to do the same, believing that the sure and only way to remedy all the evils in the world is by bringing men to submit themselves to the government of the Lord Jesus Christ.

I do here declare that I will always obey the lawful orders of my Officers, and that I will carry out to the utmost of my power all the Orders and Regulations of The Army; and further, that I will be an example of faithfulness to its principles, advance to the utmost of my ability its operations, and never allow, where I can prevent it, any injury to its interests or hindrance to its success.

And I do here and now call upon all present to witness that I enter into this undertaking and sign these Articles of War of my own free will, feeling that the love of Christ who died to save me requires from me this devotion of my life to His service for the Salvation of the whole world, and therefore wish now to be enrolled as a Soldier of The Salvation Army.

_____

_____CORPS_____ 18

.................................Corps.

....Division.

..................................18

(SINGLE)
FORM OF APPLICATION

FOR AN APPOINTMENT AS AN

OFFICER IN THE SALVATION ARMY

Name...

Address...

1. What was your AGE last birthday?............ .... What is the date of your birthday?..................................... . ...
2. What is your height?..
3. Are you free from bodily defect or disease? ....
4. What serious illnesses have you had, and when?......................
5. Have you ever had fits of any kind?...................... If so, how long ago, and what kind? ....
6. Do you consider your health good, and that you are strong enough for the work of an Officer?..................................... If not, or if you are doubtful, write a letter and explain the matter.

7. Is your doctor's certificate a full and correct statement so far as you know? ....

8. Are you, or have you ever been married? ....

---

9. When and where CONVERTED? ....
10. What other Religious Societies have you belonged to? ....
11. Were you ever a Junior Soldier? ....If so, how long ....
12. How long have you been enrolled as a SOLDIER? .... and signed Articles of War? ....
13. If you hold any office in your Corps, say what, and how long held? ....
14. Do you intend to live and die in the ranks of the Salvation Army? ....

---

15. Have you ever been an open BACKSLIDER? ....If so, how long? ....
16. Why? .... Date of your Restoration? ....
17. Are you in DEBT? ....If so, how much? .... Why? ....
18. How long owing? .... What for? ....
19. Did you ever use Intoxicating Drink? ....If so, how long is it since you entirely gave up its use? ....
20. Did you ever use Tobacco or Snuff? .... If so, how long is it since you gave up using either? ....

---

21. What UNIFORM do you wear? ....
22. How long have you worn it? ....

23. Do you agree to dress in accordance with the direction of Headquarters?..........   ....
24. Can you provide your own uniform and "List of Necessaries" before entering the Service? ....

---

25. Are you in a SITUATION?... ....If so, how long?...............
26. Nature of duties, and salary..
27. Name and address of employer? ....
28. If out, date of leaving last situation?......................... ....How long there? ....
29. Why did you leave? ....
30. Name and address of last employer?..

---

31. Can you start the SINGING?............................ ....
32. Can you play any musical instrument?.........If so, what?............
33. Is this form filled up by you?....................Can you read well at first sight?............. ....
34. Can you write SHORTHAND?.....................If so, what speed and system? ....
35. Can you speak any language other than English? ....If so, what? ....
36. Have you had any experience and success in the JUNIOR SOLDIERS' WAR? ....
37. If so, what? ....

---

38. Are you willing to sell the "WAR CRY" on Sundays?...........
39. Do you engage not to publish any books, songs, or music except for the benefit of the Salvation Army, and then only with the consent of Headquarters? ....

40. Do you promise not to engage in any trade, profession, or other money-making occupation, except for the benefit of the Salvation Army, and then only with the consent of Headquarters?..........

41. Would you be willing to go ABROAD if required?................

42. Do you promise to do your utmost to help forward the Junior Soldiers' work if accepted?...............

43. Do you pledge yourself to spend not less than nine hours every day in the active service of the Army, of which not less than three hours of each week-day shall be spent in VISITATION?........

44. Do you pledge yourself to fill up and send to Headquarters forms as to how your day is spent?...............

---

45. Have you read, and do you believe the DOCTRINES printed on the other side?...............

46. Have you read the "Orders and Regulations for Field Officers" of the Army?...............

If you have not got a copy of "Orders and Regulations," get one from Candidates' Department at once. The price to Candidates is 2s. 6d.

47. Do you pledge yourself to study and carry out and to endeavour to train others to carry out all Orders and Regulations of the Army?...............

48. Have you read the Order on page 3 of this Form as to PRESENTS and TESTIMONIALS, and do you engage to carry it out?......

49. Do you pledge yourself never to receive any sum in the form of pay beyond the amount of allowances granted under the scale which follows?...............

ALLOWANCES—From the day of arrival at his station, each officer is entitled to draw the following allowances, provided the amount remains in hand after meeting all local expenses, namely:—For Single Men: Lieutenants, 16s. weekly, and Captains, 18s; for

Single Women: Lieutenants, 12s. weekly, and Captains, 15s. weekly; Married Men 27s. per week, and 1s. per week for each child under 14 years of age; in all cases without house-rent.

50. Do you perfectly understand that no salary or allowance is guaranteed to you, and that you will have no claim against The Salvation Army, or against anyone connected therewith, on account of salary or allowances not received by you?...............

51. Have you ever APPLIED BEFORE?......... If so, when?.........
52. With what resu't?..
53. If you have ever been in the service of The Salvation Army in any position, say what? ....
54. Why did you leave?.. ....
55. Are you willing to come into TRAINING that we may see whether you have the necessary goodness and ability for an Officer in The Salvation Army, and should we conclude that you have not the necessary qualifications, do you pledge yourself to return home and work in your Corps without creating any dissatisfaction? ....
56. Will you pay your own travelling expenses if we decide to receive you in Training? ....
57. How much can you pay for your maintenance while in Training? ....
58. Can you deposit £1 so that we can provide you with a suit of Uniform when you are Commissioned?............................
59. What is the shortest NOTICE you require should we want you?...
60. Are your PARENTS willing that you should become an Officer?

61. Does anyone depend upon you for support ?.........If so, who?.....
62. To what extent? ....
63. Give your parents', or nearest living relatives', full address ....
   ....

---

64. Are you COURTING?..................If so, give name and address of the person.. ....
65. How long have you been engaged?...............What is the person's age? ....
66. What is the date of Birthday?....... . ...How long enrolled as a SOLDIER? ....
67. What Uniform does the person wear?...................... . How long worn? ....
68. What does the person do in the Corps? ....
69. Has the person applied for the work? ....
70. If not, when does the person intend doing so?........................ ..
71. Do the parents agree to the person coming into Training?....... . ...

---

72. Do you understand that you may not be allowed to marry until three years after your appointment as an Officer, and do you engage to abide by this? ....
73. If you are not courting, do you pledge yourself to abstain from anything of the kind during Training and for at least twelve months after your appointment as a Commissioned Field Officer? ....
74. Do you pledge yourself not to carry on courtship with anyone at the station to which you are at the time appointed? ....
75. Do you pledge yourself never to commence, or allow to commence, or break off anything of the sort, without first informing your

Divisional Officer, or Headquarters, of your intention to do so? ....

76. Do you pledge yourself never to marry anyone, marriage with whom would take you out of the Army altogether? ....
77. Have you read, and do you agree to carry out the following Regulations as to Courtship and Marriage? ....

> (*a*) "Officers must inform their Divisional Officer or Headquarters of their desire to enter into or break off any engagement, and no Officer is permitted to enter into or break off an engagement without the consent of his or her D.O.
>
> *b*) "Officers will not be allowed to carry on any courtship in the Town in which they are appointed; nor until twelve months after the date of their Commission.
>
> (*c*) "Headquarters cannot consent to the engagement of Male Lieutenants, until their Divisional Officer is prepared to recommend them for command of a Station as Captain.
>
> (*d*) "Before Headquarters can consent to the marriage of any Officer, the Divisional Officer must be prepared to give him three Stations as a married man.
>
> (*e*) "No Officer accepted will be allowed to marry until he or she has been at least three years in the field, except in cases of long-standing engagements before application for the work.
>
> (*f*) "No Male Officer will, under any circumstances, be allowed to marry before he is twenty-two years of age unless required by Headquarters for special service.
>
> (*g*) "Headquarters will not agree to the Marriage of any Male Officer (except under extraordinary circumstances) until twelve months after consenting to his engagement.
>
> (*h*) "Consent will not be given to the engagement of any Male Officer unless the young woman is likely to make a suitable wife for an Officer, and (if not already an Officer) is prepared to come into Training at once.
>
> (*i*) "Consent will be given to engagements between Female Officers and Soldiers, on condition that the latter are suitable

for Officers, and are willing to come into Training if called upon.

(*j*) "Consent will never be given to any engagement or marriage which would take an Officer out of the Army.

(*k*) "Every Officer must sign before marriage the Articles of Marriage, contained in the Orders and Regulations for Field Officers."

PRESENTS AND TESTIMONIALS.

1. Officers are expected to refuse utterly, and to prevent, if possible, even the proposal of any present or testimonial to them.

2. Of course, an Officer who is receiving no salary, or only part salary, may accept food or other gifts, such as are needed to meet his wants; but it is dishonourable for anyone who is receiving their salary to accept gifts of food also.

THE DOCTRINES OF THE SALVATION ARMY.

The principal Doctrines taught in the Army are as follows:

1. We believe that the Scriptures of the Old and New Testament were given by inspiration of God and that they only constitute the Divine rule of Christian faith and practice.

2. We believe there is only one God, who is infinitely perfect, the Creator, Preserver, and Governor of all things.

3. We believe there are three persons in the Godhead—the Father, the Son, and the Holy Ghost, undivided in essence, co-equal in power and glory, and the only proper object of religious worship.

4. We believe that, in the person of Jesus Christ, the Divine and human natures are united, so that He is truly and properly God, and truly and properly man.

5. We believe that our first parents were created in a state of innocency, but by their disobedience they lost their purity and happi-

ness; and that, in consequence of their fall, all men have become sinners, totally depraved, and as such are justly exposed to the wrath of God.

6. We believe that the Lord Jesus Christ has, by His suffering and death, made an atonement for the whole world, so that whosoever will may be saved.

7. We believe that repentance towards God, faith in our Lord Jesus Christ, and regeneration by the Holy Spirit, are necessary to Salvation.

8. We believe that we are justified by grace, through faith in our Lord Jesus Christ, and that he that believeth hath the witness in himself.

9. We believe the Scriptures teach that not only does continuance in the favour of God depend upon continued faith in, and obedience to, Christ, but that it is possible for those who have been truly converted to fall away and be eternally lost.

10. We believe that it is the privilege of all believers to be "wholly sanctified," and that "the whole spirit and soul and body" may "be preserved blameless unto the coming of our Lord Jesus Christ." That is to say, we believe that after conversion there remain in the heart of the believer inclinations to evil, or roots of bitterness, which, unless overpowered by Divine grace, produce actual sin; but that these evil tendencies can be entirely taken away by the Spirit of God, and the whole heart, thus cleansed from everything contrary to the will of God, or entirely sanctified, will then produce the fruit of the Spirit only. And we believe that persons thus entirely sanctified may, by the power of God, be kept unblamable and unreprovable before Him.

11. We believe in the immortality of the soul; in the resurrection of the body; in the general judgment at the end of the world; in the eternal happiness of the righteous; and in the everlasting punishment of the wicked.

DECLARATION.

I HEREBY DECLARE that I will never, on any consideration, do anything calculated to injure The Salvation Army, and especially, that I will never, without first having obtained the consent of The General, take any part in any religious services or in carrying on services held in opposition to The Army.

I PLEDGE MYSELF to make true records, daily, on the forms supplied to me, of what I do, and to confess, as far as I am concerned, and to report, as far as I may see in others, any neglect or variation from the orders or directions of The General.

I FULLY UNDERSTAND that he does not undertake to employ or to retain in the service of The Army any one who does not appear to him to be fitted for the work, or faithful and successful in it; and I solemnly pledge myself quietly to leave any Army Station to which I may be sent, without making any attempt to disturb or annoy The Army in any way, should The General desire me to do so. And I hereby discharge The Army and The General from all liability, and pledge myself to make no claim on account of any situation, property, or interest I may give up in order to secure an engagement in The Army.

I understand that The General will not be responsible in any way for any loss I may suffer in consequence of being dismissed from Training; as I am aware that the Cadets are received into Training for the very purpose of testing their suitability for the work of Salvation Army Officers.

I hereby declare that the foregoing answers appear to me to fully express the truth as to the questions put to me, and that I know of no other facts which would prevent my engagement by The General, if they were known to him.

Candidate to sign here .. ..

NOTICE TO CANDIDATES.

1. All Candidates are expected to fill up and sign this form themselves, if they can write at all.

2. You are expected to have obtained and read "Orders and Regulations for Field Officers" before you make this application.

3. Making this application does NOT imply that we can receive you as an Officer, and you are, therefore, NOT to leave your home, or give notice to leave your situation until you hear again from us.

4. If you are appointed as an Officer, or received into Training, and it is afterwards discovered that any of the questions in this form have not been truthfully answered, you will be instantly dismissed.

5. If you do not understand any question in this form, or if you do not agree to any of the requirements stated upon it, return it to Headquarters, and say so in a straightforward manner.

6. Make the question for this appointment a matter of earnest prayer, as it is the most important step you have taken since your conversion.

We must have your Photo. Please enclose it with your forms, and address them "Candidate Department," 101, Queen Victoria Street, London, E.C.

www.ingramcontent.com/pod-product-compliance
Lightning Source LLC
Chambersburg PA
CBHW022129160426
43197CB00009B/1202